E. F. Staveley

British Insects

A familiar Description of the Form, Structure, Habits, and Transformations of

Insects

E. F. Staveley

British Insects
A familiar Description of the Form, Structure, Habits, and Transformations of Insects

ISBN/EAN: 9783337186333

Printed in Europe, USA, Canada, Australia, Japan

Cover: Foto ©Andreas Hilbeck / pixelio.de

More available books at **www.hansebooks.com**

BRITISH INSECTS.

A

FAMILIAR DESCRIPTION

OF THE

FORM, STRUCTURE, HABITS, AND TRANSFORMATIONS
OF INSECTS.

BY

E. F. STAVELEY,
AUTHOR OF "BRITISH SPIDERS."

LONDON:
L. REEVE AND CO., 5 HENRIETTA STREET, COVENT GARDEN.
1871.

PREFACE.

This little work is planned on the supposition that the reader knows nothing *scientifically* of the Insect World, but that he has exercised some degree of observation on such common species as must have come before him. From this it is attempted to lead him on to a general idea of the Structure and Classification of Insects.

The main endeavour of the writer has been to induce the student to *keep ahead of the book*, which a small amount of pains in examining the very common insects chiefly described will enable him to do.

Thus, for example, after reading the first four chapters, and comparing the insects described in them with the Table of Orders (p. 60), he will find that by the time he requires the more particular tables of characters which follow the various orders, he will already be familiar with most of the characters used, and will require guidance only as to their application.

As few technical terms as possible have been employed, and, where practicable, English names have been used for the species described. This, however, is always a difficulty, from the utter absence of precision in the application of popular names; the most dissimilar insects frequently sharing one name, while one insect may be endowed with half-a-dozen "aliases" in the same county; and each one of these is the right name, and the only name, to him who employs it.

For instance—

Jan. What's got there, you?

Will. A blastnashun Straddlebob craalun about in the nammut bag.

Jan. Straddlebob! Where dedst leyarn to caal'n by that neyam?

Will. Why, what should e caal'n?—tes the right neyam, esn ut?

Jan. Right neyam, no! Why ye gurt zote vool, casn't zee tes a Dumbledore?

Will. I know tes; but vur aal that Straddlebob's zo right a neyam vorn as Dumbledore ez.

Jan. Come, I'll be deyand if I doan't laay thee a quart o' that.

Will. Done! and I'll ax meyastur to-night when I goos whoam, beet how't wool.

* * * * *

Will. I zay, Jan! I axed meyastur about that are, last night.

Jan. Well! what did ur zay?

Will. Why, a zed one neyam ez jest zo vittum vorn as tother; and he louz a ben caald Straddlebob ever zunce the island was vust meyad."*

There is a story of a young preacher who, feeling for an opinion on his sermon, elicited the compliment, "It was short." "Yes," replied the gratified orator, "I wished to avoid being tedious." "But you *were* tedious!"

The writer of the present work earnestly hopes that the attempt to be popular, yet without scientific inaccuracy, may not result in a verdict of—"Slovenly—but dry."

---

* Specimen of Isle of Wight dialect from Halliwell, in Latham's "History of the English Language."

CONTENTS.

CHAP.		PAGE
	INTRODUCTION	1
I.	THE DISTINGUISHING CHARACTERS OF INSECTS	14
II.	THE EXTERNAL STRUCTURE OF INSECTS	24
III.	THE WINGS OF INSECTS, AND THEIR CLASSIFICATION	40
IV.	THE CHANGES OF INSECTS	55
V.	COLEOPTERA	63
VI.	COLEOPTERA—(continued)	86
VII.	EUPLEXOPTERA	109
VIII.	ORTHOPTERA	113
IX.	THYSANOPTERA	123
X.	NEUROPTERA	126
XI.	TRICHOPTERA	146
XII.	HYMENOPTERA	152
XIII.	HYMENOPTERA—(continued)—TEREBRANTIA	155

CHAP.	PAGE
XIV. TEREBRANTIA—(*continued*)	169
XV. HYMENOPTERA—(*continued*)—ACULEATA	187
XVI. ACULEATA—(*continued*)	203
XVII. ACULEATA—(*continued*)	213
XVIII. ACULEATA—(*continued*)	221
XIX. ACULEATA—(*continued*)	231
XX. LEPIDOPTERA	255
XXI. LEPIDOPTERA—(*continued*)	268
XXII. LEPIDOPTERA—(*continued*)—LARVÆ	276
XXIII. HOMOPTERA	296
XXIV. HETEROPTERA	315
XXV. APHANIPTERA	329
XXVI. DIPTERA	332

GLOSSARY	380
INDEX TO FAMILIES, GENERA, ETC.	382
GENERAL INDEX	387

DESCRIPTION OF PLATES.

PLATE I.
COLEOPTERA.
SECTION I.—PENTAMERA.

Adephaga.

		PAGE
Fig. 1. Cicindela campestris (*Tiger Beetle*)	. .	66
,, 2. Carabus violaceus	. .	67

Hydradephaga.

| ,, 3. Acilius sulcatus ♂ . . . . | . . | 69 |
| ,, 4. Gyrinus natator (*Whirligig Beetle*) . | . . | 70 |

Necrophaga.

| ,, 5. Silpha quadripunctata (*Burying Beetle*) . | . | 73 |

Brachelytra.

| ,, 6. Goërius oleus (*Devil's Coach-horse*) | . . | 75 |

PLATE II.
COLEOPTERA—(*continued*).
PENTAMERA—(*continued*).

Clavicornes.

| Fig. 1. Byrrhus pilula (*Pill Beetle*). 1, *a*. Profile of head . . . . | . | 77 |

Lamellicornes.

| ,, 2. Geotrupes stercorarius (*Common Dung Beetle*, or *Dumbledor*) . . . . . | . | 78 |
| ,, 3. Melolontha vulgaris ♀ (*Cockchafer*) | . . | 79 |

Macrosterni.

| ,, 4. Elater (Athoüs) vittatus (*Skipjack*) | . . | 80 |

Aprosterni.

| ,, 5. Lampyris noctiluca ♂ (*Glowworm*), male. 5, *a*. Profile of head . . . . . | . | 81 |
| ,, 6. Ditto ditto female | . . | 81 |

PLATE III.

COLEOPTERA—(continued).
Pentamera—(continued).
Aprosterni—(continued).

Fig. 1. Telephorus fusca (*Sailor*) 81
 ,, 2. Anobium striatum (*Death-watch*) . . . 84

Section II.—Heteromera.
Trachelia.

 ,, 3. Pyrochroa rubens (*Cardinal Beetle*) . . 86
 ,, 4. Meloe proscarabæus (*Oil Beetle*) . . . 87

Section III.—Tetramera.
Rhyncophora.

 ,, 5. Phyllobius argentatus (*Weevil*). 5, *a*. Profile of head 89

Longicornes.

 ,, 6. Clytus arietis (*Wasp Beetle*) 93

PLATE IV.

EUPLEXOPTERA.

Fig. 1. Forficula auricularia (*Earwig*), with wings expanded 110

ORTHOPTERA.

 ,, 2. Blatta lapponica (*Small Cockroach*) . . 115
 ,, 3. Acheta domestica (*House Cricket*) . . . 116
 ,, 4. Phasgonura (Gryllus) viridissima (*Great Green Grasshopper*) 120
 ,, 5. Locusta flavipes 122

THYSANOPTERA.

 ,, 6. Phælothrips cerealis (*Thrips*) 123

DESCRIPTION OF PLATES.   xi

PLATE V.
NEUROPTERA.
                                                        PAGE
Fig. 1. Ephemera vulgata (*Mayfly*) 135
 ,, 2. Panorpa communis ♂ (*Scorpion-fly*) . . 140
 ,, 3. Hemerobius perla (*Lacefly*) 139
 ,, 4. Sialis lutaria 140
 ,, 5. Raphidia ophiopsis (*Snakefly*) . . . 141

TRICHOPTERA.
 ,, 6. Phryganea grandis (*Caddis-fly* or *Water-moth*) 146

PLATE VI.
HYMENOPTERA.
SECTION I.—TEREBRANTIA.
Serrifera.
Fig. 1. Tenthredo zonata (*Sawfly*) 158
Terebellifera.
 ,, 2. Sirex Gigas (*Woodborer*), less than natural size 168
Spiculifera.
 ,, 3. Cynips lignicola (now *Kollari*) (*Gallfly*) . 177
 ,, 4. Ophion luteus (*Yellow Ophion*). 4, a. Side
        view 180
 ,, 5. Chalcis flavipes 182
Tubulifera.
 ,, 6. Chrysis ignita (*Ruby-tail*) 184

PLATE VII.
HYMENOPTERA—(*continued*).
SECTION II.—ACULEATA.
Heterogyna (Ants).
Fig. 1. Formica flava (*Yellow Ant*) 188
 ,, 2. Mutilla Europæa (*Solitary Ant*) . . . 201

PLATE VII.—(*continued*).

Fossores (*Sand and Wood-wasps*).

		PAGE
Fig. 3. Pompilus exaltatus	. . . . .	206
,, 4. Ammophila sabulosa	. . . . .	206
,, 5. Tachytes pompiliformis	. . . . .	207
,, 6. Mellinus arvensis. 6, *a*. Head of ditto, profile		208

PLATE VIII.

HYMENOPTERA—(*continued*).

ACULEATA—(*continued*).

Fossores—(*continued*).

Fig. 1. Crabro vagus	. . . . . .	210
,, 2. Pemphredon lugubris	. . . . .	211

Diploptera (*True Wasps*).

,, 3. Eumenes coarctata (*Solitary Wasp*)	. .	213
,, 4. Odynerus antilope (*Solitary Wasp*)	. .	214
,, 5. Vespa vulgaris (*Common Wasp*) (*Social*). 5, *a*. Face of ditto	. . . . .	215
,, 6. Vespa Norvegica (*Social Wasp*). 6, *a*. Face of ditto	. . . . . .	220

PLATE IX.

HYMENOPTERA—(*continued*).

ACULEATA—(*continued*).

Andrenidæ (*Short-tongued Bees*).

Fig. 1. Sphecodes rufescens	. . . . .	227
,, 2. Halictus morio ♂	. . . . .	227
,, 3. Andrena fulva ♀	. . . . .	229

Apidæ (*Long-tongued Bees*).

,, 4. Nomada sexfasciata ♀	. . . . .	234
,, 5. Osmia bicornis ♀	. . . . .	236
,, 6. Bombus terrestris ♀	. . . . .	243

DESCRIPTION OF PLATES. xiii

PLATE X.

LEPIDOPTERA.

SECTION I.—RHOPALOCERA.

	PAGE
Fig. 1. Gonepteryx rhamni ♂ (*Sulphur Butterfly*)	260
„ 2. Hipparchia janira ♀ (*Meadow-brown Butterfly*)	262
„ 3. Polyommatus alexis ♂ (*Common Blue Butterfly*)	264

SECTION II.—HETEROCERA.

Sphingina.

„ 4. Chærocampa porcellas (*Small Elephant Hawk-Moth*) 269

Bombycina.

„ 5. Pygæra bucephala (*Buff-tipped Moth*) . . 270

Noctuina.

„ 6. Gonoptera libatrix 272

PLATE XI.

LEPIDOPTERA—(*continued*).

HETEROCERA—(*continued*).

Geometrina.
(Not represented here.)

Pyralidina.

Fig. 1. Botys urticata (*Small Magpie Moth*) . . 273
„ 2. Hypena proboscidalis (*Snout Moth*) . . 273

Tortricina.

„ 3. Xanthosetia zygæna 274

Tineina.

„ 4. Cemiostoma laburnella 275

Pterophorina.

„ 5. Pterophorus pentadactylus (*Strawberry Plume Moth*) 275

Alucitina.

„ 6. Alucita polydactyla (*Twenty-plume Moth*) . 275

PLATE XII.
HOMOPTERA.
Section I.—Trimera.

		PAGE
Fig. 1.	Cicada anglica	300
„ 2.	Anthophora spumaria (*Cuckoo-spit Insect*)	302
„ 3.	Cercopis sanguinolenta	302
„ 4.	Membracis cornuta. 4, *a*. Ditto, side view	302

Section II.—Dimera.

„ 5.	Aphis Rosæ (*Rose Aphis*). 5, *a*. Side view, natural size	303
„ 6.	Aleyrodes chelidonii. 6, *a*. Ditto, three times natural size, to show position of wings	308

PLATE XIII.
HETEROPTERA.
Section I.—Hydrocorisa.

Fig. 1.	Notonecta glauca (*Water Boatman*)	316
„ 2.	Nepa cinerea (*Water Scorpion*)	318

Section II.—Aurocorisa.

„ 3.	Gerris paludum	320
„ 4.	Capsus spissicornis	322
„ 5.	Lygæus equestris	323
„ 6.	Pentatoma rufipes	324

PLATE XIV.
DIPTERA.
Section I.—Proboscidea.
Nemocera.
Culicidæ.

Fig. 1.	Culex pipiens ♂ (*Common Gnat*). 1, *a*. Head of ditto ♂ (antennæ truncated). 1, *b*. Ditto ♀ (antennæ truncated)	347

Brachycera.
Stratiomidæ.

„ 2.	Stratiomys chameleon (*Soldier-fly*)	354
„ 3.	Sargus cuprarius. 3, *a*. Antenna	354

Tabanidæ.

„ 4.	Tabanus autumnalis (*Horsefly*). 4, *a*. Antenna	355

DESCRIPTION OF PLATES.

PLATE XIV.—(continued).
Asilidæ.
Fig. 5. Asilus crabroniformis 356
Leptidæ.
„ 6. Leptis Scolopacea. 6, a. Antenna . . 357

PLATE XV.
DIPTERA—(continued).
PROBOSCIDEA—(continued).
Brachycera—(continued).
Bombylidæ.
Fig. 1. Bombylius major (Beefly) 357
Empidæ.
„ 2. Empis tessellata 358
Syrphidæ.
„ 3. Eristalis tenax (Dronefly) . . . 47, 360
„ 4. Syrphus pyrastri. 4, a. Antenna . . . 361
„ 5. Melithneptus menastri 362
Conopidæ.
„ 6. Conops rufipes 363

PLATE XVI.
DIPTERA—(continued).
PROBOSCIDEA—(continued).
Brachycera—(continued).
Muscidæ.
Fig. 1. Bucentes geniculatus 364
„ 2. Stomoxys calcitrans (Stable-fly) . . . 364
„ 3. Musca domestica (Housefly). 3, a. Antenna . 365
„ 4. Sepsis cynipsea 368
Œstridæ.
„ 5. Œstrus (or Cephalemyia) ovis (Gadfly). 5, a. Antenna 369
SECTION II.—EPROBOSCIDÆ.
„ 6. Melophagus ovinus (Sheeptick) . . . 372

LIST OF VIGNETTES.

EGYPTIAN HAWK-HEADED SCARABÆUS THRUSTING FORWARD THE DISK OF THE SUN.
From a Carved Stone in the British Museum. Page 63.

TITHONOUS.
From an antique gem. Page 113.

LION AND BEE.
Roman Mithraical gem from the *Gemmæ Antiquæ* of Agostino. Page 152.

DEATH'S HEAD BUTTERFLY AND URN.
From a Neo-Platonian gem, signifying the Immortality of the Soul. Page 255.

Or, *Ephemera?* signifying the shortness of life. See page 10, and note, page 136.

DEUS MYIAGRUS (the God of Flies). See page 2, note.

ERRATA.

Page 38, title of cut, *read* "Naucoris cimicoides."
„ 103, the line "SUBSECTION I.—ADEPHAGA," should *precede* the three lines above.
„ 115, last line, *for* "Lapponia," *read* "Lapponica."

INSECTS.

INTRODUCTION.

As the object of this work is, not to teach Entomology, but to lead the reader in the first steps towards that science, by increasing the interest with which he may be disposed to regard the tribe of insects, it may not be altogether out of place to devote a few pages to their connexion with the history and superstitions of past ages.

They are a small people, but they have no small work to accomplish in the world. They are a small people, but they hold no inconsiderable place in the history of mankind. In our own day there are whole tracts of country where their dominion cannot be overthrown by man, and from whence he is driven by them. There was a time when a mighty king was shaken in his purpose " by reason of the swarm of flies," and there was a time when nations bowed down before the Lord of Flies.

The worship of the Fly, or rather of the Fly-destroyer personified, is said to have commenced in Egypt. From Egypt the Caphtorim carried it into Palestine, and there we find their descendants, the Philistines of Ekron, worshipping the Fly-god under the name of Baal-zebub.*

---

* 2 Kings, i. 2.

By the Phœnicians this worship was introduced into Tyre, Sidon, and Babylon, and from these three great centres of commerce and civilization it spread into other parts of the world.

In Greece the origin, according to tradition, of this worship was, that Hercules, being tormented during the Olympic rites by hosts of flies, offered a sacrifice to Zeus in order to be rid of them. The sacrifice was accepted, and the flies removed beyond the boundary of the River Alpheus. From this time the great Zeus was known at Olympia by the surname of 'Απόμυιος (Apomyius)—"*driving away the flies,*"—and the annual sacrifice of a bull to Zeus Apomyius* at the Olympic games, is said to have been performed with the result of dispersing the hosts of flies, which were the torment of those rites; whilst the Elians were unremitting in the like worship, by which they deprecated the infliction of those swarms of flies, which they believed to bring with them pestilence and disease. At the festival of Athena at Aliphera, the Hero Myiagrus, or Myioides (μυίαγρος, that is, the *fly-catcher*) was invoked as the protector against flies.

The Romans also had their Deus Myiagrus, and into the Temple of Hercules, at Rome, flies were not permitted to enter.

Coming nearer to our own day, we read of the same or a similar worship as prevalent amongst the Hottentots, who adore " as a benign deity, a certain insect, peculiar, it is said, to the Hottentot countries. This animal is of

---

* A representation of Zeus Apomyius, or the Deus Myiagrus, on an ancient gem, will be found figured at the head of the chapter on Diptera. The face of the god is given in the figure of the fly.

the dimensions of a child's little finger, the back is green, and the belly speckled with white and red. It is provided with two wings, and on its head with two horns. To this little winged deity, whenever they set eyes on it, they render the highest tokens of veneration; and if it honours a *Kraal* (a village) with a visit, the inhabitants assemble about it in transports of devotion, as if the LORD OF THE UNIVERSE was come among them. They sing and dance round it while it stays, troop after troop throwing to it the powder of *Bachu*, with which they cover at the same time the whole of the kraal, the tops of their cottages, and everything without doors. They likewise kill two fat sheep as a thank-offering for this high honour. It is impossible to drive out of a Hottentot's head that the arrival of this insect to a kraal brings favour and prosperity to the inhabitants."*

That this worship should have obtained so widely, will not seem wonderful, when we recal the historical evidences of the power of these little creatures, and remember that under the polytheistic system of religion, not only were the beneficent powers of nature adored, but the agents prejudicial to man were personified, and became the objects of deprecation. Hence, it could hardly fail that creatures so powerful for evil as to be the means of devastating and rendering uninhabitable whole tracts of country, should find a place amongst the fear-inspiring gods of the heathen.

Thus, too, it may easily be conceived that while the Israelites of old were rejoicing over the messengers of their All-Powerful Protector, that plague, which took its

---

* Kolben's "Present State of the Cape of Good Hope," vol. i., quoted in Parkhurst's Greek Lexicon, under "Beelzebub."

rank amongst such miseries as pestilence, murrain, hail with fire, and bereavement, had to the Egyptians yet another horror added when in it they found their great deity subject to the bidding of the leader of their oppressed slaves.

Very different from the place held by the fly is that occupied by a representative of another order of insects, namely, the Bee. Seldom, or perhaps never, actually the object of adoration, it finds its place in the symbolism and amongst the superstitions of all times and countries.

It is as

> "Creatures that by a rule in Nature teach
> The art of order to a peopled kingdom"

that they are found amongst the hieroglyphs of Egypt, the symbol of royalty being, according to Horapollo, a reed (or sceptre) followed by a bee; denoting the people obedient to a king.*

It may have been in the same sense that it was adopted as a badge by the ancient kings of France, as, for instance, by Childeric, on the opening of whose tomb in St. Denis, above 300 golden Bees, which had formed the decoration of his robe, were found; whilst it is known that Louis XII. and Henri IV. sometimes used these emblems instead of *fleurs de lys*. Upon this it is conjectured that the *fleur de lys* was a corruption of the

---

* Hence too perhaps arose the superstition prevalent among the Greeks and Romans that the sudden appearance of a swarm of bees was inauspicious and ominous of slavery. Virgil, in his 4th Georgic, says of them that—
> "Not Egypt, India, Media, more
> With servile awe their idol king adore."— DRYDEN.

Mr. Sharpe, the great Egyptologist, denies that the bee and sceptre in the hieroglyphs conveyed this meaning.

figure of a bee, the three upper leaves representing the body and wings of a bee with the head downwards, while the lower parts of the leaves took the place of the head and legs, &c.

The Great Napoleon, who while changing the established order of things, never missed an opportunity of showing that he knew full well the value attaching to the *prestige* of antiquity, replaced the dishonoured *fleur de lys* by the imperial and more ancient badge of the bee; and his coronation robe, probably in imitation of that of Childeric, was "*semé*" with golden bees.

As a symbol of plenty or fecundity, the bee, or its produce, occurs on all sides. The "Land flowing with milk and honey" of the Bible, is the most familiar instance of this, and the combination is found elsewhere.

Thus in the Hindoo Mythology, in which Maya, the Mother of the World and of the Sea of Milk, or primitive matter, holds so high a place, we find the bee also bearing a part among the symbols of fecundity. Cama (*Love*), is represented as a child-god, supported on a quiver, from which issues Yotma (*strength*), under the form of a lion, the group resting on a bee. Yotma is also represented under the form of a compound being, with a head half bull half lion, the wings of an eagle and the body of a serpent. From the yawning mouth of this being proceeds another deity—Prakriti (*goodness*), in the form of a cow, accompanied by a swarm of bees. But there is one curious figure which exceeds these in interest. Maya, the Creatrix, holds her child, the infant God of Love, in her arms; behind him is his quiver, and in his hand a bow of sugar-cane strung with bees. Possibly the sting is the point of this figured epigram, which at least reminds us of the answer of Venus to young Love's complaint,—

> "Oh, mother, I am dead!
> An ugly snake, they call a bee,
> O see it swell! hath murdered me.
>
> "Venus with smiles replied, 'Oh sir,
> Does a bee's sting make all this stir?
> Think what pains then attend those darts
> Wherewith thou still art wounding hearts,'" &c.

The great blue bee also appears in the Hindoo Mythology, reposing on a lotus, and sacred to Vishnu, the second person (or *preserver*) of the Trimurrti, or Trinity.*

The bee is found on the coins of those parts of Greece in which the ancient and beneficent god Aristæus, son of Apollo and Cyrene, was worshipped. He taught men to keep bees, and the medals of Athens, of Ceos, and other places, bear this insect as his attribute. It occurs also on the coins of Ephesus, the city worshipping the great goddess of all fertility and abundance, whose symbol was a bee.

Again, the bee occurs in the representations of the mithraical worship of Persia, as afterwards adopted in Rome, where the principle of fertility or production is combined with that of strength, under the figure of a lion with a bee at the mouth, forcibly reminding us of the Hindoo use of the same symbols. A very beautiful ancient gem with this subject is figured by Agostino (*Gemmæ Antiquæ*), in which, according to him, allusion is made to the riddle of Samson, "Out of the strong came forth the sweet."†

---

* For figures of the above Hindoo representations, see the "*Nouvelle Galerie Mythologique, par J. D. Guigniaut.*" *Paris*, 1850.

† A woodcut of this gem will be found at the head of the first chapter on Hymenoptera.

In Greece bees were recognised as omens of future eloquence, and the stories are well known of the swarming of bees upon the lips of Pindar and of Plato, who—

> "Did shed
> Sweet words like dropping honey."

And the title of "The Attic Bee" was bestowed upon Xenophon.*

Later, Antonius, a Greek monk (of the eighth? twelfth? century), who formed two Books of Sentences collected from the rich field of the writings of the early Christian Fathers, was surnamed "Melissa," or the Bee; and Leo Allatius (keeper of the Vatican Library in the seventeenth century), gave to the illustrious men of his own time the collective name of *Apes Urbanæ*.

It is not easy to account for some of the modern superstitions which attach to bees.

The county of Kent is rich in these; there, if the bees swarm upon a dead tree, the result is a death in the family of their owner; and so strong is the feeling upon this subject, that care is taken to avert such a misfortune by cutting down any dead tree before the time of the swarming of the bees. In the same county the intimate relation between the hive and the household is also shown by a curious custom which prevails of waking up the bees by knocking on the hive, to tell them when a death occurs in the family. In Brittany (and in Cornwall?) they tie a small piece of black stuff to the beehives at the time of a death, and a piece of red in the case of a marriage; without this the bees would never thrive. In the district of Quimperlé, if the hives

---

* Contrast with these our "Wasp of Twickenham."

have been robbed, the bee-keeper immediately gives them up, there being an old Breton proverb, "No luck after the robber."*

In Ireland, bees are considered "the luckiest things at all," and an unfortunate house and unsuccessful dairy have been known to go right from the moment of the arrival of swarming bees.

These "smallest among fowls" have found a place even in heraldry. They were in the family arms of Urban VIII., in whose pontificate Allatius wrote his *Apes Urbanæ*, and in England "three bees volant, *azure*, on a ground, *or*," are borne by the family of *Bye*, formerly the Saxon, and still the Dutch name for the bee.

The Ant, an insect of the same order as that to which the bee belongs, is the subject of a curious superstition in Ceylon, which is quoted by Messrs. Kirby and Spence from Knox's "Ceylon." There is a species of black ant there which "bites desperately, as bad as if a man were burnt by a coal of fire; but they are of a noble nature, and will not begin unless you disturb them. Formerly these ants went to ask a wife of the *Noya*, a venomous and noble kind of snake; and because they had such a high spirit to dare to offer to be related to such a generous creature, they had this virtue bestowed upon them that they should sting after this manner. And if they had obtained a wife of the Noya, they should have had the privilege to sting full as bad as he."

Like the bee, the ant is present in representations of the god Mithras, and Plutarch tells that it was used in divination.

We will turn now from these tribes of ruling, ruled,

---

* Nesquét a chunche, varlearch ar laër.

and provident creatures, with their much-lauded virtues, to the joyous, musical, sun-loving tribes of grasshoppers and cicadas—" harmless creatures, nourished upon dews," as was once fondly believed, and whose song is to the peasant a harbinger of fair weather and a plentiful harvest. And here again we will quote from the Introduction to Entomology—

".... They were addressed by the most endearing epithets, and were regarded as all but divine. One bard entreats the shepherds to spare the innoxious tettix, that nightingale of the nymphs, and to make those mischievous birds the thrush and blackbird their prey. 'Sweet prophet of the summer,' says Anacreon, addressing this insect, 'the Muses love thee, Phœbus himself loves thee, and has given thee a shrill song; old age does not wear thee; thou art wise, earth-born, musical, impassive, without blood: thou art almost like a god.'"

Our authors go on to suggest that the τέττιξ of the Greeks must have been more musical than those of other countries, which have been "execrated for the deafening din that they produce;" but there is as great variety in musical taste as in the quality of music, and among English poets we find one attributing the "sweet music" of the woods to the chorus of lark, linnet, throstle, nightingale, and grasshopper; while another writes of "screaming grasshoppers," which "fill everye eare with noyse." That the *cicada* itself entertained little doubt of its musical powers was proved in a contest between Eunomus and Ariston at the Pythian games, when, one of the strings of the cithara of Eunomus being broken, a cicada perched upon the instrument, supplied the deficiency, and won the day for him.

We cannot, in a chapter devoted to such associations,

pass from the grasshopper and his associates without one word of him whom "only"—

"Cruel immortality consumes:"

Who dwelt—

"In presence of immortal youth,
Immortal age beside immortal youth."

Tithonous appears to have been seldom made the subject of representation in ancient art; but there is a curious gem which represents him "undergoing his metamorphosis," of which an engraving is placed at the head of the chapter on Orthoptera.*

To speak of the butterfly as connected with the *superstitions* of past ages would be an injustice. It stands forward amongst the corrupted myths of the ancients, a beautiful example of pure symbolism, and Psyche ($\psi v \chi \acute{\eta}$) or the soul, is almost constantly, in the later periods of ancient art, to be recognised by her butterfly wings.

First, the grovelling life of this world—crawling and feeding upon the earth; then the deathlike sleep—silent and motionless; then the breaking forth free, beautiful, and winged—surely it is not wonderful that to the poetical Grecian mind, man, living, dead, immortal, was pictured here.

And thus we find it in a thousand representations. On the lips of Plato, preacher of the immortality of the soul, rests a butterfly;† and the symbol was introduced into early Christian Art by his descendants, the Neo-

---

* It will be observed that insects of two orders have here been mixed, but though separated in science, as musicians they are closely connected, and it is sometimes not easy to ascertain to which animal some notices of the ancients, on this point, are to be referred.

† Or sometimes butterfly's wings are on his head.

Platonians, to whom is attributed an engraved gem in which a butterfly hovers over a death's-head.*

Perhaps a more interesting series of examples of the application of this symbol could hardly be found than is presented by the bas-relief on a Roman sarcophagus, described by Maury,† in which are set forth the course of man's destiny, his creation, the imparting to him a living soul; his life and sufferings; his death, or the parting of soul and body; and the transportation of the disembodied spirit.

The subject begins with the creation of man, and the reader must be prepared for a little confusion, entailed by Prometheus bearing a double character as creator and as the prototype of man.

Prometheus is represented seated, holding the finished man, the work of his own hands, upon his knee. Before him stands Minerva, in the act of placing *the Butterfly* on the head of the newly-created ("and man became a living soul"); whilst near this group are seen *Terra* (the Earth, from whence all men come), and *Cupid* and *Psyche*, who, embracing each other, set forth the union of soul and body. Above, the fates are busy; Clotho winds the thread of man's life upon a spindle, while Lachesis traces his horoscope upon a globe.

The next scene represents the sufferings of Prometheus (as man), and Deucalion and Pyrrha, types of the perpetuation of the race of mankind, are present; but, whilst the race subsists, the individual passes away, and the next figure is of the lifeless body extended before Atropos, who sits with the book of destiny open, whilst Love, in

---

* Figured at the head of the chapter on Lepidoptera.
† "Nouvelle Galerie Mythologique."

the character of the Angel of Death, watches the *butterfly* which is escaping from the body. Terra, present at the birth, is here present at the death, as if to take back to herself the mortal remains; while Mercury, the soul-bearer, is seen transporting the figure of Psyche, or the soul—a female, with butterfly's wings—to the regions of the blest.

Amongst our own less observant and less poetical countrymen, we may perhaps refer the naming of these insects to a feeling of superstition, or a state of mind akin to that described by Bishop Taylor, when "every bush is a wild beast, and every shadow is a ghost, and every glowworm is a dead man's candle, and every lantern is a spirit;" and accept Messrs. Kirby and Spence's suggestion that it is from the old notion that the dead fly about at night in search of light, that in the north and west of England the nocturnal moths which fly into the candles are called *saules* (souls), as in Germany they are "ghosts;" while the Italians believe the fireflies to be spirits arisen from the graves, and avoid them in terror.

It is gratifying to turn from the contemplation of superstitious cowardice to the example of valour tempered by mercy, given by our British Ajax Telamon, who, "when grown as mad as any hare (For he had sought each place with care, And found his Queen was missing)"—

"He next upon a glowworm light,
(You must suppose it now was night,)
Which, for her hinder part was bright,
  He took to be a devil :
And furiously doth her assail,
For carrying fire in her tail ;
He thrashed her rough coat with his flail ;
The mad king feared no evil.

INTRODUCTION.

> 'Oh,' quoth the glowworm, 'hold thy hand—
> Thou puissant king of fairy land.
> Thy mighty strokes who may withstand?
>   Hold, or of life despair I :'
> Together then herself doth roll,
> And tumbling down into a hole,
> She seemed as black as any cole,
>   Which vext away the fairy."
>   DRAYTON'S *Nymphidia*.

To enter upon any account of the *Scarabæus*, or Sacred Beetle of the Egyptians, would be but to burden the reader with matter with which he must be already familiar, and its place in the symbolism of Egypt, where, bearing its orb-like burthen, it represents the vivifying power of the sun, is too well known to require more than this passing notice. It would seem, however, that veneration for the Beetle tribe is not confined to that ancient nation, as it is said that in Sweden there is a belief that any one who shall place an overturned cockchafer on his legs will have three sins remitted to him. It is to be hoped that cockchafers are plentiful in Sweden.

Many more are the details which might be collected of the place held by insects in history and in literature, but the present chapter has already over-passed all reasonable limit, and we must proceed to the more deeply interesting facts laid open by an examination of the objects themselves.

CHAPTER I.

ON THE DISTINGUISHING CHARACTERS OF INSECTS.

THE name "Insect" is in common parlance applied very indiscriminately to whole classes of animals which have little in common, except the smallness of their size. Flies, earthworms, tadpoles—creatures farther removed from each other in their intimate structure than are the horse, the shark, and the eagle, are sometimes confounded together, and called "Insects."

Nor is this all; we have read that "flies are bred from worms;" that certain large moths are "a kind of little birds;" that murex* is "a genus of insects belonging to the order Vermes Testacea," and is "of the snail kind!" Nay, the writer once heard a lady reply to some remark upon a mouse, that she did "not like *any* insects!"

Ignorance such as this is perhaps now rare; yet it is doubtful whether, even amongst those who do know that a mouse and a tadpole are not insects, there are not many persons who would be sorely puzzled to tell in what the difference consists, and who would be surprised at the assertion that a tadpole or a snake is more nearly related to a horse or an eagle, than to any wriggling grub in the waters or creeping worm upon the earth; and that the whole tribe of flying insects, whether large or small,

---

* "Enyclopædia Britannica."

are in their nature and construction farther removed from the birds than these are from the lion and the tiger; nay, that that which is the apparent link between these "fowls" (to use the inclusive term of an old writer), namely, the power of flight, is attained by organs which are absolutely without any relation but that of their function; and which consequently form no true link as regards the structure and constitution of the two animals, the bird and the insect.

The object then of the present chapter is to show the characters by which insects are distinguished from animals of other classes. To do this it will be necessary to lay before the reader a slight sketch of some of the leading characters of both one and the other.

Throughout the animal kingdom we find several plans, as it were, or systems; even as the animal kingdom itself is one system among many others in creation. It appears as if the Creator had confined within these plans or systems such variations of detail as were essential to attain all the ends which He had in view, and what these ends were we may in a great measure ascertain by the study of nature herself, learning from this the work which is done, and the variety of life and enjoyment with which our world is filled, by the multiplication of living beings under a diversity of form, habit, and character.

The whole of the animal kingdom is divided by science into two great classes, namely, the Vertebrata, or animals possessing a spinal column, and the Invertebrata, or animals which are without this. It must be noted, however, that while the class Invertebrata contains many systems or groups, as those of the insects, the worms, the "shellfish," and others, Vertebrata contains but a single such system or group, within which are only such differences as are produced by variation in the details of parts. In this class are included man, beasts, birds, reptiles, and fishes.

It is now time to support the assertion that a tadpole comes nearer in its nature to a horse, or an eagle, or a mullet, or a man, than to the grub of a water-beetle born and bred in the same stagnant pool with itself; and to do this a few words must be given first to those vertebrate animals, and next to that section of the Invertebrata to which insects belong.

The tadpole is a young and undeveloped frog, and can, of course, be spoken of only as a frog. Now the frog, the horse, the bird, the fish, the man, agree in these respects,—they all possess an internal bony and jointed framework, or skeleton, composed of living tissues, nourished throughout life by bloodvessels, and growing with the growth of the animal. To this framework the muscles are attached. The principal parts of this skeleton are the spinal column or backbone, with the ribs, the skull, and the bones of four limbs. All these parts are not however found in all vertebrate animals, and indeed the backbone seems to be the only part of the skeleton which is never wanting. Thus, for instance, the frog has no ribs, the snake has no limbs, and there is a fish which has no skull. Again, these parts, when present, are in various animals variously modified, enabling each to fill its own place in creation; and in an examination of these modifications we perceive the connexion existing among these animals under the greatest diversity of form and habit.

To take the four limbs, for example. In man they form two legs and two arms; in the ape four arms; in the horse four legs. In the tortoise we might hesitate whether to call them hands or feet, while in the turtle they are fins. The bird offers yet another variety, its limbs consisting of two legs and two wings. Now a glance at a skeleton will show how small a difference of development in the proportions and direction of the bones makes the difference between the foot and the hand in man. It takes more study to find that the foot of the horse is the foot of the man, saving that only one enormous toe with a proportionate nail (the hoof) is developed, the rudiments of but two others appearing on dissection, while the rest are altogether atrophied. This will appear less startling when the development is traced next in the two-toed animals ("cloven-footed"), as the cow, the three-toed rhinoceros, the pig with its four toes —two big and two little, the cat with its five-clawed toes, one of which takes a direction separate from the others like the human thumb.

These examples must suffice: it would be out of place in this work to trace the variety of development which, from the same system of bones, produces the human hand, the wing of a bird, and the hundred-fingered fin of the skate; or again, to trace the atrophy of parts by which in reptiles the limb dwindles down to a mere indication —as in the slow worm—to be altogether lost in the true snakes. No line can be drawn between the highest and the lowest of the vertebrata which shall separate them from each other so clearly as it will be shown that they are separated from all invertebrate animals.

Passing from the framework of the body to the organs by which the vital functions are performed, we find in all the vertebrata a nervous system originating and centring in the brain, whence, by ramifications from the spinal chord, the whole body is supplied with nerves, the mysterious vehicles of communication between mind and body —nerves of motion carrying the orders from the mind to the body, nerves of sensation carrying information from the body to the mind, with much more of the abstrusest nature. In all this the brain appears to be the centre of life, and if communication be cut off between the brain and any member, that member becomes useless.

The nutrition, or building up of the body of an animal (vertebrate or invertebrate) is a compound operation, consisting first, of the collection of material; secondly, of its preparation; thirdly, of its application; and fourthly, of its reparation when deteriorated. The collection of material is simply the process of feeding. Its preparation is that of digestion, by which is elaborated out of various substances a fluid containing the ingredients necessary to the nourishment of the body. The third and fourth processes, namely, the application and reparation of this building material, *the blood,* are dependent on *circulation* and *respiration.*

In the vertebrata these take place as follows:— The blood, elaborated by digestion from the food, is committed to the heart, which first sends it to the lungs to take in a supply of oxygen from the air which they contain, and then receives it back again to send it forth on its journey through the body. This it performs, at first through the channel of the arteries, visiting every part, and in each depositing some of its constituents. One organ robs a portion of it of one substance, another of another; each organ, attending to its own business, chooses, probably through the operation of its nerves, its required material. Secretions are formed, tissues are constructed, bones, muscles, fat, receive their appropriate food; laboratories unnumbered, working without cessa-

DISTINGUISHING CHARACTERS OF INSECTS. 19

tion, are without cessation supplied with material for their mysterious operations.

Through the arteries this stream of life, impelled by the repeated action of the heart, continues to pour; but at each point a portion of the stream, deprived of its oxygen and of other component parts, becomes debased and loses its value and power: by another series of channels therefore, the veins, it flows back towards the heart, receiving from the stomach on its way fresh nutritive material, again is sent to the lungs to gather from the air contained in them the oxygen by which its life-giving powers are renewed, and again is propelled into the arteries to recommence the circuit of the body.

Now, leaving untouched all other anatomical details and physiological phenomena, let us compare what has been said of the vertebrata with the facts which we find in the invertebrata, using in this comparison, in order to shorten and simplify the chapter, only the order of *insects* from among the invertebrata in which this order forms a perfect group.

Insects are without any internal skeleton at all. The body is supported by an external more or less hard and jointed case, which forms the covering of the body, and to which the muscles are attached, as in the vertebrata to the internal skeleton. This case, in fact, answers the purposes of both skin and skeleton.

The limbs of a perfectly-developed insect consist of six jointed legs, neither less nor more; certain four-legged butterflies being merely instances of aborted limbs, while in the many-legged caterpillars the extra "legs," as they are called, are sucker-like and jointless processes of the skin.

As in the vertebrata, so in insects, there is great

diversity of development in the parts, and great modification of form to suit the needs of various modes of life. But, while we may sometimes trace a curious resemblance to certain of the vertebrata in the functions, and even in the external forms of these members, their fundamental differences are as great as ever, and the adherence of each to the principles of the separate plan or system to which it belongs only becomes the more evident when the same end is attained under various systems by means always in accordance with those systems. Thus the burrowing mole-cricket has a flattened, hand-like fore-leg, which forcibly reminds us of the mole ; the grasshopper has the large and springing thigh of a frog ; the water beetle has fin-like legs : but in each of these, we find that it is but a change in the proportion of the parts which makes the difference between the legs of the cricket, the grasshopper, the beetle, even as we found before that the limbs of the vertebrata have one series of parts variously modified.*

Besides these six legs, the perfect insect is furnished with two pairs of wings. The wings of a bird are, as has been said, composed of the same bones as those which form the forelegs, or arms, of other vertebrata, only under a different proportion and development of parts.

The wings of an insect are, on the contrary, an appendage of the breathing apparatus; with this they are closely connected, and it may be supposed that it receives assistance from them in the performance of its functions.†

---

* This will be illustrated by figures in the second chapter.

† That a communication exists between the lungs and the wing bones in birds, and that the acts of respiration and of flight affect each other in these also, is a farther proof of analogy in function under a different plan in structure.

DISTINGUISHING CHARACTERS OF INSECTS. 21

Taking the nervous system next, we find an important difference between the vertebrate animals and insects. In insects, instead of one nervous centre or brain, a series of nerve-knots, called ganglions, communicating with each other, and yet acting apparently with some degree of independence, send off the supply of nerves required by the body. Thus, while a vertebrate animal dies at and below the point at which the connexion with the brain is cut off, an insect may be cut into several pieces, and to all appearance each may, for a considerable time, show signs of vitality. Thus a headless insect may be seen to walk; and a dragonfly, accidentally divided into three parts, the head, the thorax, and the abdomen, has for days kept up considerable action in the separated parts, the wings fluttering violently on any attempt to confine them, and the abdomen wriggling when touched.

The circulating and breathing systems are next to be compared.

Until recently (*i.e.*, within the last fifty years) insects were believed to be without any heart or circulating system whatever, although a certain motion of fluids had been observed before that time. Now, however, it has been shown that a long muscular vessel, which is in fact a sort of compound heart, or series of heart-valves terminating in a large artery, runs from the end of the body into the head. Here this vessel branches off, and although from the extreme delicacy of its minute offshoots these have been traced but a little way, there seems reason to believe that a system of arteries and veins exists on the same principle as in the vertebrata.

The hearts of the vertebrata and of insects are not more unlike than are the organs of respiration. In the place of the lungs, two large spongy bodies, full of air-cells, to which, in the vertebrata, the blood is brought for aëration by the heart; in insects the air is carried in tubes, or air passages, to every part of the body. Down each side of the insect runs a large air tube, communicating by short tubes running out of it with the breathing holes, or spiracles, which lie along each side of the abdomen. From these two main passages shoot little clusters of smaller tubes, which ramify again and again, until their minute branches are to be found in every part of the body. Thus, while the vertebrate animal inhales only through the double passage terminating in the mouth (or, as in fish and reptiles, in the gills), the insect breathes through a series of openings in its abdomen; and the air, instead of being carried to the chambers in which the blood visits it, is carried to the blood in every part of the body.

It now remains to define the especial character of the true insect, and to show in what it differs from other animals not separated from it by barriers of so decided a nature as those just mentioned.

The name insect is now much more restricted in its use than it formerly was. Spiders, centipedes, scorpions, woodlice, shrimps, and even lobsters, have been included under the term, but are now considered as belonging to other orders. The true insect, as at present received, is an animal arriving at maturity through a series of moults, or metamorphoses. It is without internal skeleton, having the body enclosed in a jointed covering, and is composed of three principal parts—head, thorax, and abdomen; the head bearing antennæ, the thorax bearing six jointed legs, and (with certain exceptions) four wings.

The reader will bear in mind that this definition applies to the perfect insect only, and that caterpillars, footless fly-grubs, and the like, are but young and imperfect insects, bearing indeed something of the same relation to those full grown as does the tadpole to the frog.

They are no arbitrary characters which thus separate insects from nearly allied races. Thus, the spider, excluded from the list of insects by its eight legs and its head and thorax being in one mass instead of two, is also separated from them by far more important internal characters; as, for instance, the possession of true lungs. Thus also the many-legged woodlouse, shrimp, &c., with the spider, are more widely separated from the six-legged insect by the absence of metamorphosis, than by any difference in the number of their limbs.

These non-changing animals attain the perfect state merely by increasing in size and in the perfection, sometimes also in the number, of their parts. From time to time they cast their skin, as it becomes too small to contain them, but they undergo no essential change of form or character after their exclusion from the egg.

Insects, on the contrary, as has been said, undergo a series of "metamorphoses" or changes, more or less complete, before arriving at the perfect or winged state. It is true that these metamorphoses are but developments, and that the chrysalis, for instance, is not *changed into* a butterfly, but that it is itself a *butterfly in a husk;* still the difference between the perfect and the imperfect insect is, as a rule, so great, and the stages of development are so marked, that the word metamorphosis may fairly be applied.

CHAPTER II.

ON THE EXTERNAL STRUCTURE OF INSECTS.

IT is not proposed to enter into any elaborate account of the anatomy of insects, and many things usually considered to be of the alphabet of Entomology, will be omitted in this chapter. Much of this however, laborious and uninteresting if studied first in descriptions, is acquired gradually, without trouble, and indeed almost unconsciously, as the various species of insects come, one by one, under the notice of the student. A short account therefore of such parts only as it is absolutely necessary to know by name will be given, with familiar examples, in the hope that the reader will, if possible, examine for himself every insect named, and observe for himself many things not noted here.

The name Insect (as also the Greek ἔντομα, *entoma*, whence "Entomology") is given on account of the cut, or divided character of the body. The body consists of a series of joints, or rings, called segments. These are soldered together so as to be with difficulty distinguishable in some parts of the body, but are usually very evident in the abdomen. The perfect insect is divided into three principal parts: the head, formed of one, or, as some say, of more than one segment; the thorax, to which the wings and legs are attached, and which is composed of three segments closely united; and the abdomen, in which the number of segments varies, nine, or possibly ten being the highest number found.

EXTERNAL STRUCTURE OF INSECTS. 25

The organs principally to be noticed in the head are the eyes, the antennæ or "horns," and the mouth.

The eyes of insects are of two kinds, simple and compound. The reader must often have observed the large convex brown eyes of the common house-fly, and those lustrous bodies which form so conspicuous a feature in the dragonfly, and the glorious golden or ruby eyes of the delicate lace-fly. He may also have observed that these eyes are immovable, and consequently, the motion of the head being very limited, would be of little use either in the avoidance of danger or in the pursuit of prey, were they constructed like our own and able to see in one direction only. That the vision of the fly is *not* so limited will be amply proved by a few attempts to "get on the blind side" of one. Approach him from above, from below, from before, from behind, from either side, or from round the corner, he perceives and avoids the danger. How is this? The large eye which we observe on either side of the head is in fact a *cluster* of eyes, or, to speak more properly, is a *compound* eye. Looking closely at this eye in one of the large insects, the reader will observe the surface marked out into hexagons (fig. 1). Each of these is the surface of a true eye, and the hexagonal form is such as a number of cylindrical or conical eyes would naturally assume if pressed together. It will be seen that such a cluster of eyes, if arranged so as to form a semi-globular surface, would be able to see all objects on one side of an insect, whilst the fellow eye could receive impressions from objects on the other side. These eyes then are so arranged.

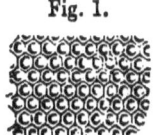

Fig. 1.

Small portion of eye highly magnified.

It has been said above that the compound eye is

composed of many cylindrical, or rather *conical* eyes (which for convenience sake we will call eyelets), whereas the common idea of an eye is that it is an organ of globular form. It would require more space than can be afforded here to describe the structure of the cylindrical eyelet; it must suffice to say that in it are represented nearly all the principal parts which exist in the human eye, even to the iris and pupil, although together they form a long slender cone instead of a globe* (fig. 2). This striking difference finds its explanation when we consider that this form, by reducing the rays of light which can possibly reach the retina to such only as fall directly upon it, prevents the confusion which would arise from the reception by each eyelet of images from all sides. It may be observed here that some insects (*e.g.* the butterfly) possess as many as 34,650 of these eyelets.

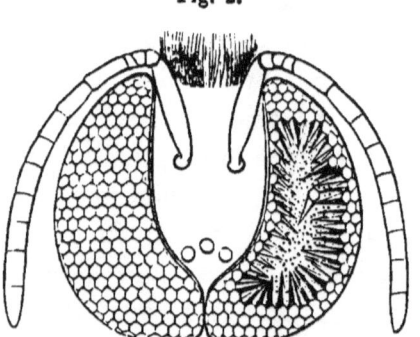

Fig. 2.

Eyes of Hive Bee (Male).

Very curious observations have been made upon the connexion which exists between the size and position of the eyes, and the flight of the insect. Thus the dragonflies and butterflies, alike remarkable for the freedom and extent of their flight, have large and convex eyes, so placed that the field of vision must be very great. Others again, as the bees, which have long and narrow

---

* In this figure (partly borrowed from Dr. Carpenter) the lenses are made large out of proportion, in order the better to show their form.

EXTERNAL STRUCTURE OF INSECTS. 27

eyes, with upward and downward but little lateral scope, have an irregular flight, usually directed up, down, or forward, but very little from side to side. The field of vision of the grasshoppers, &c., is still more limited, and their flight is short and hesitating. In the earwigs, which fly little and crawl much, the eyes are small, and are placed on the top of their flat heads.

It has also been observed that in the eyes of some insects (as the dragonfly) the eyelets are by no means uniform in size, the surfaces of the upper being considerably larger than those of the lower eyelets. The inference from this appears to be, that there is some variation in the length of the sight—the eyelets turned earthwards being probably shorter sighted than those which are on the look out for birds of prey and other aërial enemies.*

Besides these compound eyes most insects have also a set of semi-globular *simple eyes*, or "*stemmata*" or "*ocelli.*" These are usually three in number and placed in a triangle upon the forehead (see fig. 6). They may be easily observed in the bee, wasp, and dragonfly. It is supposed that they are intended for the perception of near objects only.

What has been written applies only to *perfect* insects. The *larvæ* of insects (*i.e.*, insects in an early stage, as caterpillars, &c.) never possess compound eyes, but (with the exception of such as from their subterranean or other habits require no eyes, and therefore have none) have always one or more pairs of simple eyes, resembling the stemmata already mentioned. Where several of these exist,

---

* For much more on the subject of insects' eyes the reader is referred to some interesting papers by Mr. Parson "On the Discoveries of Müller and others," in the "Magazine of Nat. Hist." for 1831; and to "The Honey Bee, its Natural History, Habits, &c," by James Samuelson and Dr. J. B. Hicks: Van Voorst.

as in the larva of Dyticus, a large water beetle, which has five or six on each side, they are still simple, being independent of each other, and having separate optic nerves, whereas in compound eyes one large optic nerve or ganglion sends branches to all the eyelets.*

The antennæ of insects have long perplexed naturalists, who have in turn ascribed to them every known sense but that of sight—smell, touch, taste,—even hearing. It is not, however, even yet ascertained to what sense they belong. They are evidently of the greatest importance to the insect in a variety of ways, aiding it in its perceptions and guiding it in its actions, and this in matters so various as to suggest that it may be the organ of some sense or senses to which we have nothing corresponding, or of the combination of which, at least, our limited experience gives us no means of forming an idea.

The forms of antennæ to be observed are very various, and, in some instances, exceedingly beautiful. Some are thread-like, others clubbed, and others feathered—all these are found in moths and butterflies: others again are like strings of beads, or are toothed like a comb, or terminate in a fan, as may be seen in beetles. In one species of these last the antenna is about four times as long as the body, while in certain species of flies it consists principally of a little globe, with a curved bristle sticking out of it, like the reaping-hook by which Daniel O'Rourke held on to the moon.

These organs are so valuable in the determination of genera, &c., that the young student should from the begin-

---

* These simple eyes are found in the spiders, woodlice, centipedes, &c.; none but the true insects (*i.e.*, those which undergo metamorphosis) possessing the compound eyes.

EXTERNAL STRUCTURE OF INSECTS. 29

ning accustom himself to observe their character in every insect which comes before him, always remembering, however, that there is often much difference between the antenna of the male and female, even in the same species.

The mouths of insects, as might be expected, afford a most interesting variety of structure, being adapted not only to the various nature of the food proper for the different species, but also to their various modes of life. Thus, among the architectural species, parts of the mouth are so modified as to act as spades, trowels, &c.; to the upholsterers they are scissors; in the chase they seize and hold the prey, while the warlike tribes find in them powerful weapons of offence.

The principal parts of the mouth are six; the upper and lower lips, and two pairs of jaws, those of each pair acting upon each other, from side to side. An idea of the arrangement of the mouth may be formed from the accompanying diagram, in which A represents the upper lip, or labium; B, the lower, or labrum; c c, the upper pair of jaws, or mandibles; d d, the lower pair, or maxillæ. (See also fig. 6, p. 32.)

A	
c	c
d	d
B	

Insects are divided primarily into biting and sucking insects, and while the parts just named are easily recognised in the first division, they are in the second so differently developed and modified as to be hardly traceable, except by such a process as that described in chap. i., for tracing the relationship between the horse's foreleg, the bird's wing, and the arm of a man.

The mouth of a beetle affords an excellent example of all these parts, as they are found in biting insects Fig. 3 represents the top of the head of a Tiger-beetle (Pl. I., fig. 1), and shows the situation of the mandibles, which are very large in this insect—a formidable pair of pincers when extended as in the figure (*b*), and lying quite across each other when closed (*a*). These organs are even larger and more conspicuous in the male of the Stag-beetle, whilst in most other beetles they are much smaller and less powerful.

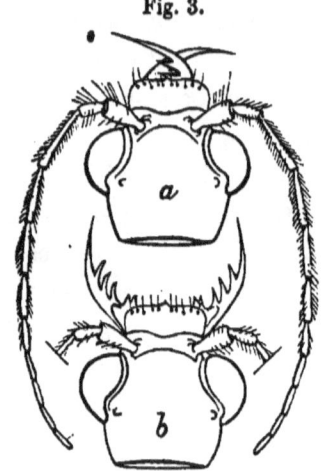

Fig. 3.

a Head of Tiger-beetle, (magnified) with jaws closed.
b Ditto, with jaws open.

The second pair of jaws, or the maxillæ (fig. 4), are more complicated and delicate. The principal parts of the maxilla are a kind of blade (*a*), fringed with hairs, and an antenna-like feeler or palpus (*b*), called the *maxillary palpus*. In the Tiger-beetle and some other predaceous beetles there is a second appendage (*c*) resembling a less developed palpus.

The upper lip is a horny plate, without appendages. The lower lip, or "labium" (fig. 5), is very

Fig. 4. Fig. 5.

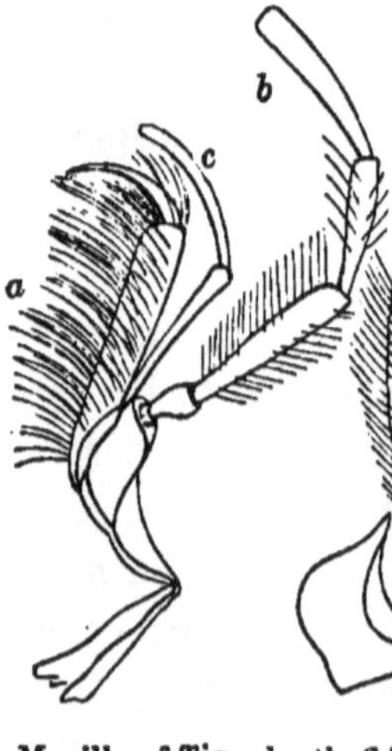

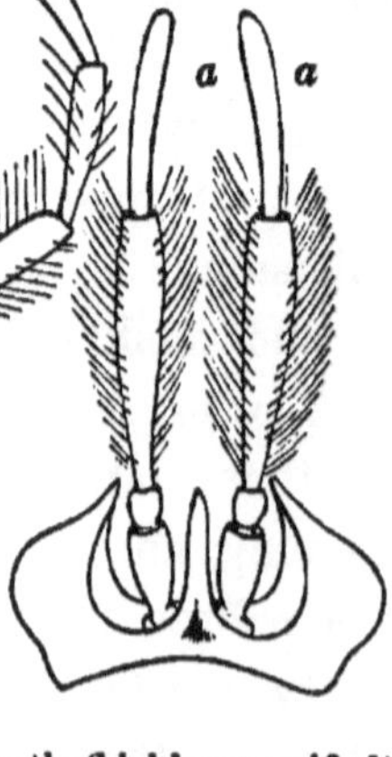

4. Maxilla of Tiger-beetle (highly magnified).
5. Labium of Tiger-beetle.

EXTERNAL STRUCTURE OF INSECTS. 31

different in different orders of insects. It is composed of several pieces, more or less developed, and both these separate pieces and the whole organ have been variously named by various authors. It is generally composed of a basal horny plate, succeeded by a second horny or membranous plate, something like a lip, or by a prolonged fleshy tongue-like organ; and always bears a pair of palpi, called labial palpi (fig. 5, *a a*, and 6, *f f*). Sometimes the whole organ is called the labium or lip, sometimes only the second part is so called, the first being called mentum or chin. Sometimes the whole is called "lip," "tongue," "proboscis," and so on. In the present work the whole organ will generally be called by the most usual name of labium, while the English words "lip," or "tongue," will be applied according to the form, whether it be, as in beetles, for instance, a lip-like plate, or, as in bees, a fleshy projectile tongue-like instrument. In the dragon-fly and the grasshopper the lining of the lip is free, and forms an internal tongue something like our own.

In the first five orders of biting insects—namely, those which contain the beetles, earwigs, grasshoppers, dragonflies, and caddis-flies,* no important variation occurs in the character of the parts of the mouth as described above. In the sixth order, however, containing the bees and their relations, we come to the first remarkable change in the form of these parts, though they are still to be recognised with ease. The peculiarities

---

* The caddis is included in this list for the sake of uniformity, but in fact, these insects, living but a short time in the perfect state, and requiring little or no food, have the mouth in a very rudimentary and undeveloped state.

of development being most conspicuous in the bee itself, this shall be taken as an example.

The upper lip, or labrum (fig. 6, b), and the mandibles, or upper jaws (c c) of the bee resemble those of other biting insects. The mandibles are of various forms in the several genera (as will be shown in chap. xix., but are always strong, horny, biting jaws). The maxillæ or lower jaws (d d), however, of the bee, entirely lose their jaw-like character, and become long, thin, membranous plates (always bearing the maxillary palpi,*) and fulfil the office, when drawn together, *of a sheath to the tongue.* This tongue, or ligula (g), is a long, slender, hairy organ, growing on a fleshy base, and is, in fact, a prolongation of the "*labium*" (e e), the fleshy base being sometimes called the "*mentum*," and on each side of the tongue (as the organ is here called, having altogether lost its lip-like character) the two labial palpi (f f) are found. Besides the palpi, the tongue is furnished with two slender filaments, called paraglossæ (παρὰ, near, γλῶσσα, the tongue), which are found also in some other biting insects. The tongue of the bee,

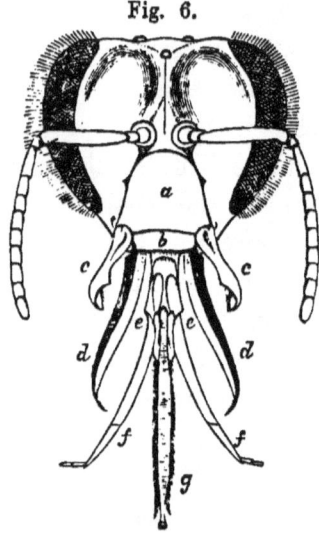

Fig. 6.

Face of neuter hive-bee, magnified. *a*, clypeus; *b*, labrum; *c*, mandibles; *d*, maxillæ; *e*, labium; *f*, labial palpi; *g*, ligula of the labium.

---

* The maxillary palpi and paraglossæ are not shown in this figure, but may be seen in that of Anthophora retusa, in the twelfth chapter.

EXTERNAL STRUCTURE OF INSECTS.

enclosed in its sheath and folded close under the breast, will be easily seen in the first hive or humble bee the reader may examine; and, if the bee has not been so long killed as to have become stiff, all the parts of the organ can be opened out and displayed by placing a needle below the tongue and drawing it forward.

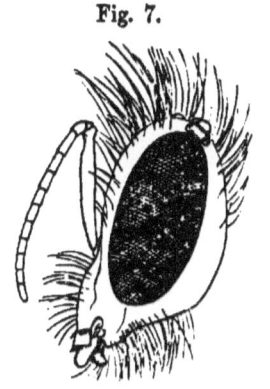

Fig. 7.

Profile of Neuter Hive Bee, with the tongue folded out of sight.

In some bees the labium is comparatively short, and in different genera cleft or acute, as will be described in a later chapter.

With the bees we come to the end of the first division of insects; namely, those which have biting mandibles. In the second division, containing the insects with *sucking* mouths, and without biting jaws, the whole structure of the mouth appears widely different. The stinging proboscis of the gnat, the fleshy blunt trunk of the housefly, the long slender tongue of the butterfly, all these display variation of structure.

The "tongue" in these insects is composed of some or other of the parts already mentioned, recognisable though greatly altered from the biting type, and occasionally so soldered together, or transformed in figure, that nothing but a careful analysis can reduce them to a uniform plan. This will not be attempted here, but it would be well for the reader to aim at tracing the connexion existing in the organs of the various insects he dissects from time to time.

In the butterflies and moths a long tubular proboscis is found, which coils up under the mouth when at rest (fig. 8—1, 2). This is a development of the maxillæ, other parts of the mouth, excepting the lower lip, being almost undeveloped. The under lip is furnished with a pair of large palpi, thickly clothed with hair.

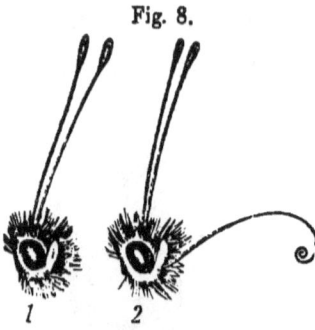

Fig. 8.

Profile head of Moth.

Next in order come the aphis, cicada, &c., and the water-boatmen, bugs, &c. The proboscis of these is a fine but sometimes very hard tube, containing four hair-like lancets. These lancets wound the surface of plant or animal, the juices of which are then sucked up through the tube. In this case the tube is formed by the labium, the four lancets representing the maxillæ and mandibles.

In the flea the mouth is a sucking apparatus with a pair of serrated lancets; but the parts, though closing upon each other when at rest, differ from those of the two preceding orders in being free and independent of each other.

The two-winged flies—such as the housefly, the gnat, and the Daddylonglegs, present some variety in the form of mouth; but in all a series of lancets and a sucking tongue are the main features. This tongue is an exceedingly beautiful object when magnified, and is very easily examined in the large bee-like 'Drone-fly,' described in the following chapter.

The appendages to the Thorax of insects are, as has been said, the legs and wings. The wings having been used as the basis of the classification of insects, will form the subject of a separate chapter.

The legs and wings are attached to the parts of the

EXTERNAL STRUCTURE OF INSECTS. 35

thorax as follows:—the first segment bears the first pair of legs; the second segment bears the second pair of legs and the first pair of wings; the third segment bears the third pair of legs and the second pair of wings.

The legs of insects are six in number. These are adapted to several modes of progression, as walking, leaping, burrowing; while they serve other purposes also, the fore legs being prehensile in some species, the hind legs adapted to carry burdens in others, and so on. The principal parts of the leg are, the coxa; trochanter; femur; tibia, and tarsus (see fig. 9). The coxa (*a*) is a large and flat joint hinged to the body. It is very conspicuous in the large water-beetle, *Dyticus*. The trochanter (*b*) is the next, and a very small joint. In the hind leg of some insects, *e.g.* saw-flies, it is formed of two pieces instead of one. The next joint is the femur (*c*) or thigh—the large and usually thick joint which stands out horizontally from the insect's body. Next is the *tibia* (*d*), or shank, usually of about the same length as the femur, but thinner; and lastly the tarsus (*e*), which is composed of a series of joints, terminating in a clawed foot. The joints in the tarsus in different insects are from one to five in number.

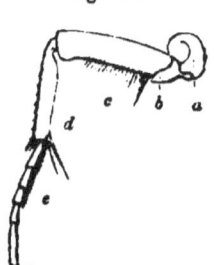

Fig. 9.

Middle leg of Dyticus marginalis.
a. coxa.
b. trochanter.
c femur.
d. tibia.
e. tarsus.

It is in the fore and hind legs that we find the most striking variations of development for special ends; the following figures are intended for comparison with figure 9, in order to give the reader an idea of the manner in which the parts are modified. Fig. 9 shows an ordinary form of insect's leg. Fig. 10

is the hind leg or oar of the common water-boatman. This leg is a true oar, and little more. It is elongated and strongly fringed with hairs, only two joints

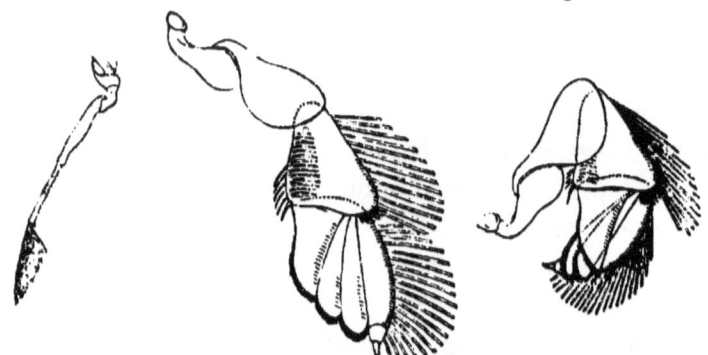

Fig. 10. Hind leg of Notonecta.
Fig. 11. Hind leg of Gyrinus extended, greatly magnified.
Fig. 12. The same contracted.

are developed in the tarsus, and the claws are generally wanting.

Fig. 11 is the hind leg of the little shiny black "whirligig beetle," also a denizen of the waters, where it is rendered conspicuous by the marvellous rapidity of its evolutions. The wonderful little living paddles by which these motions are made deserve close attention, and far exceed in beauty the oars of the Notonecta. They are quite flat, the femur, tibia, and joints of the tarsus being composed of horny plates beautifully articulated together. The femur and tibia are triangular. The three upper joints of the tarsi are excessively dilated on the inner side, so that when expanded they form, with the crescent-shaped fourth and fifth joints, a thin semicircular disk. The limb in this state opposes to the water these broad flat horny plates. When contracted (fig. 12), the tarsal joints fold over each other like the vanes of a fan, and may consequently be drawn through the water with little

EXTERNAL STRUCTURE OF INSECTS. 37

resistance. The limb is, like most swimming legs, beautifully fringed with hairs.

In figure 13 is seen a leg fitted for leaping, the leg of the common grasshopper; and the thick and muscular thigh, the strong but slender and spurred tibia, and the firmly knit but supple joints, all point to the action for which this limb is adapted. Besides this, the great length of the hind or leaping legs as compared with the two other pairs should be remembered, and the leaping leg of the grasshopper will be seen to be as good an example of peculiar development for a special purpose as the swimming legs lately described.

Fig. 13.

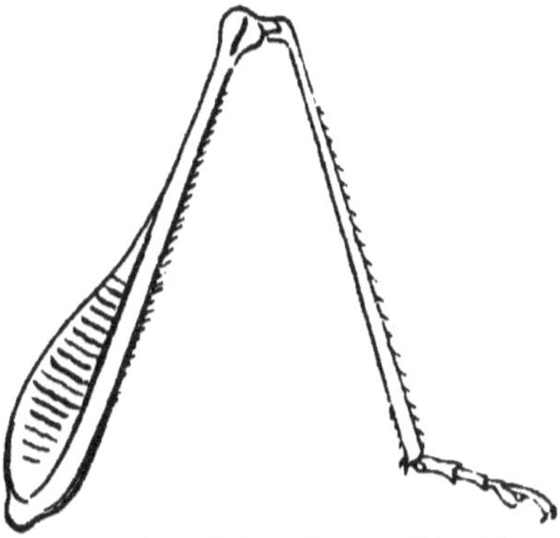

Hind leg of Grasshopper (*Acrida viridissima*).

From this we turn to the *fore leg* of a near relation of the Grasshopper, but an insect of far other habits Figure 14 represents the burrowing mole-like *hand* of the mole cricket. In this curious instrument, as in the paddle of the gyrinus, the tibia and tarsus are unusually broad and flat, and so arranged as to be capable of fitting close to each other and to the thigh. The tibia is deeply cut into finger-like lobes, to which it owes its hand-like appearance, and like the broad short

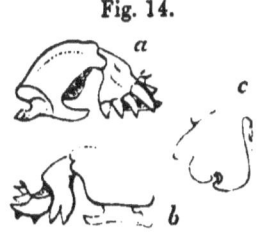

Fig. 14.

a. Fore leg of Mole Cricket, seen from outside.
b. Ditto from inside.
c. Coxa of fore leg.

hand of the mole forms a most admirable implement for burrowing.

Figure 15 displays a pair of unmistakeable nippers, and woe betide the luckless shrimp or larva which finds itself between the forceps of the water-scorpion. Of the *raptorial* character of these arms there can be little doubt.

Fig. 15.

Head, thorax, and foreleg of *Naucori cinicaule*, magnified, one claw closed.

It would require too much space to describe here the wonderful foreleg of the water-beetle *Dyticus*, with tarsus dilated into a disk, covered with the most exquisite little membranous suckers;—the hind leg of the hivebee, furnished on one side with a basket in which to carry home the stores of pollen collected from flower to flower, and on the other, with rows of combs for use in its manipulations within the hive;—or many another curious and beautiful illustration of the variety to be found in the legs of perfect insects alone. In those of various larvæ there are yet other forms, but, as a general rule, these are more simple than in the imago.

The feet of insects are curious and beautiful. The commonest form is of two claws with one, two, or three soft pads; but the pads are often wanting, and sometimes one or both claws. Further description of them is unnecessary here, as they are objects which the student will find no difficulty in examining for himself.

The abdomen has little to describe of external organs, the principal being the various ovipositors or instruments for the placing of the eggs, which will be described later, and the spiracles or breathing holes spoken of page 22. These are sometimes exceedingly

EXTERNAL STRUCTURE OF INSECTS.

beautiful, and perhaps the most strikingly so is the fringe-guarded spiracle of the large water-beetle (Dyticus). The reader may easily prepare this for examination by simply cutting through the skin under the wing-cases, removing the part containing the spiracle and washing it well with water and a camel-hair brush. The two pairs of breathing holes nearest to the tail are the largest and most beautiful.

CHAPTER III.

ON THE WINGS OF INSECTS, AND ON THEIR CLASSIFICATION.

THE young student of Entomology will perhaps be agreeably surprised to find that characters so obvious as those presented by the wings of insects, are set before him as the basis of the classification of the tribe. It will at once occur to him that if the four network wings of the dragonfly, the two membranous wings of the house-fly, the down-covered wings of the butterfly, are sufficient to point out the orders to which these insects belong, then it cannot be very difficult to take the first step in Entomology—that of determining to what order any insect belongs.

That he may not, however, suppose this character to be chosen merely as a means of sorting insects not essentially allied, he must, for the present, take for granted that of which he will soon perceive the truth—namely, that a certain character of wing is found to correspond with more important characters in the general structure and habits of insects, and that the orders thus formed are in fact natural groups.

The beetles stand first in most arrangements of insects, with them, therefore, the description should commence. But let it be understood that by "beetles" are not meant "black beetles," and that "black beetles" are not beetles.

"What then is a beetle?"

The name of the order to which beetles belong will partly answer this question. *Coleoptera*, from κολεὸς (Koleos), a sheath, πτερὸν (Pteron), a wing. A beetle may be described as a four-winged insect, whose first pair of wings, being thickened to a horny or leathery consistence, form a covering or "sheath" to the hind pair. This is the most conspicuous character of the order, and one by which nearly all beetles may at once be recognised. The common cockchafer will serve as a familiar example, for there are few persons who do not know this insect. There are even perhaps few who have not watched it raise its brown wing-cases, and, spreading them apart, unfold the large and beautiful wings, delicate glistening membranes, extended and supported by strong nerves, which are to them what the spars of a ship are to the sails. But no ship's spars are so jointed as these nerves—no sails so reefed as these membranes; the captain of no ship could fling out his sails, let the wind blow which way it will, and, helmless, trust them to bear him to his haven. Yet this can our little beetle do. And this he does, let it be observed, with a single pair of wings, whilst the other pair, thickened, and utterly useless in flight, unless indeed they serve to guide it, might seem to be even a hindrance to his motions. Now, certainly, we should not have expected that *beetles*, perhaps on the whole the most ponderous of flying insects, should thus have been deprived of half their support by such a modification of the very organs of locomotion as the conversion of one pair of wings into a sheath for the other.

How then is this loss compensated? In the beetles the hind wings (which in most other orders of insects are considerably the smaller) are much larger and more fully developed than the fore wings, so much so indeed as to be able to do the double share of work which falls upon them. Then, to ensure the safety of the wing—a wing exceeding in size the sheath which should protect it—it is furnished with a double set of joints, which enable it to be folded and packed closely beneath the wing-case. The wing folds longitudinally, and at the same time a hinge-like joint in the longitudinal nerves, about one-third from the tip, allows it to be turned inwards and shortened (fig. 16). The process of unfolding this may be easily seen by watching a ladybird, cockchafer, or other slow-moving beetle.

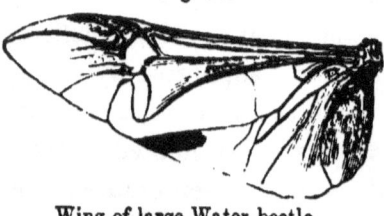

Fig. 16.

Wing of large Water-beetle.

Some beetles are without the second fold in the wing, the wing being wider, but not longer than the wing-case.

Other beetles again, such as the well known "Devil's coachhorse," have the wing-case so short as not to cover a third part of the abdomen, yet so perfect is the folding of the underwing that it is in most cases entirely covered by the wing-case.

For figures of Order I., Coleoptera, see Plates I., II., III.

Those beetles which have short wing-cases are followed naturally by the earwigs, which resemble them in this particular, while they are distinguished from them by the pincer-like termination of the body, and more especially by the form, veining, and folding of the wings, which also are not entirely covered by the very short wing-case, *the exposed part being protected by a thickening of the membrane.* The wing of the earwig is very broad, the outline being rather more than the quarter of a circle. The veins radiate from a point in the thickened part of the membrane, and the wing is packed first by being closed together like a fan and then transversely folded *in two places*. (fig. 17). From this complicated double folding is derived the name of the order to which the earwigs belong, viz., EUPLEXOPTERA (εὖ, well, πλεκτός, folded ; πτερὸν, wing).

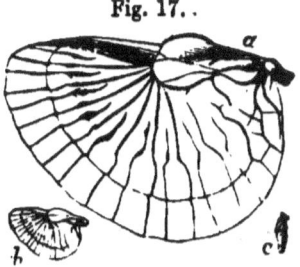

Fig. 17..

a. Wing of Earwig magnified.
b.   do.   do.   natural size.
c.   do.   do.   closed.

For figures of Order II., Euplexoptera, see Plate IV., fig. 1.

To the earwig, the grasshopper, cricket, locust, and cockroach (or blackbeetle of our kitchen) succeed. Resembling the earwig in the fan-like folding of the hind wing, they differ from it in having no transverse folding (fig. 18), and from this character of the wing is derived the name of the order under which these insects are ranged; namely, Orthoptera, or straight - winged (ὀρθὸς, straight ; πτερὸν, a wing.) The fore wings, although

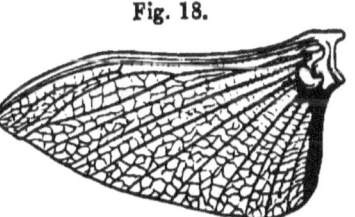

Fig. 18.

Wing of Grasshopper (*Acrida viridissima*).

much thickened, are less thick and horny than those of either the beetles or the earwig, and are useful in flight.

And here we come upon a most curious little apparatus. The merry chirp of the house cricket and of the grasshopper are amongst our most familiar sounds, yet few inquire the nature of the instrument by which the little creature produces its pleasant music. This, the primitive violin—with bow, string, and sounding-board,—is to be found in the fore wings.

This instrument is most conspicuous in the crickets. It consists of a clear space in the wing-cases, or fore-wing, consisting of a tense membrane enclosed by strong and prominent nervures; near this lies a strong nerve or ridge, with a toothed, file-like surface. This file (*the bow*), most prominent on the upper surface of the wing which underlies, and on the under surface of that which overlaps, plays, when the wings are rubbed together, upon the raised ribs, causing a strong vibration in the drum-like membrane, or sounding-board, beside them, and thus producing the sound.

Figure 19 shows the drum and file (or sounding-board

Fig. 19.

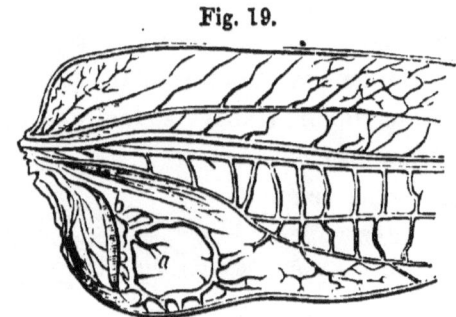

Base of under side of wing-case of green Grasshopper (*Acrida viridissima*) magnified.

and bow) in the left wing of the green grasshopper. Fig. 20 shows the instrument in the right wing (*A*) of another species, Acrida brachelytra, in which one of the strings crosses the sounding-board. *B* is the left wing, on which in this species is the file or bow. *C* is the file or bow more highly magnified. In the common house cricket the sounding-board is divided by nervures into several areas of various sizes and shapes, and the sound is supposed to be influenced by this circumstance.

WINGS OF INSECTS, AND THEIR CLASSIFICATION. 45

It has been found that the sound may be produced artificially in dead specimens by rubbing the wings together.*

Fig. 20.

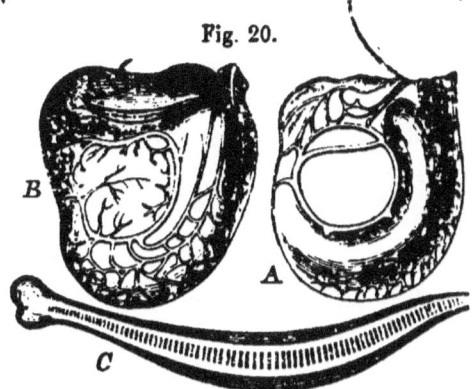

A. Upper surface of right wing-case of Acrida brachelytra.
B. Under surface of left wing-case of do.
C. File more highly magnified.

For figures of Order III., Orthoptera, see Plate IV., figs. 2, 3, 4, 5.

From the insects whose encased and folded wings have been described, we turn to the dragonfly, with four large, strong, ever-expanded wings, which bear the insect forward, backward, or from side to side with equal ease, and with a swiftness far beyond that of almost any pursuing enemy.

These wings, with those of the smaller dragonflies, of the delicate golden-eyed lace-fly, and others, are chiefly characterized by their numerous nerves, which, intersecting their whole surface, form a kind of fine network of small squarish meshes. The insects with these wings, and some others in which the network is not so perfect, belong to the order Neuroptera (νεῦρος, nerve, πτερὸν, wing). The wings are always four in number, and

---

*. These figures, with some others used in this work, are taken from Todd and Bowman's "Cyclopædia of Anatomy and Physiology."

in some, but not in all cases, the four are of equal size.

For figures of Order IV., Neuroptera, see Plate V., figs. 1, 2, 3, 4, 5.

The large, hairy, moth-like wing of the Trichoptera (τριχιων, hairy; πτερὸν, wing) are the sign of another order. It is not difficult to mistake the large brownish or drab-coloured caddis fly, or "water-moth," as it is called by anglers, for a true moth. The downy wings of this are, however, clothed with simple hairs, instead of the dust-like scales with which the wings of the moth are covered, and very slight attention will prevent any mistake.

For a figure of Trichoptera, see Plate V., fig. 6.

Next come the bees, wasps, ants, sawflies, and others. These have four clear wings with fewer nerves than those of the dragonfly, and which form fewer, or sometimes hardly any, meshes; indeed, in the case of some minute genera the wings are altogether without veins. As then the name of a former order, Neuroptera, was taken from the predominance of the nerves, so in the naming of this, as *the membrane* predominates, the order to which these insects belong is called Hymenoptera (ὑμην, a membrane; πτερὸν, a wing).

In the wings of this order we find mechanism as unexpected as that in the wing of the cricket, though of a different nature, and as an example we will take the common hive bee, so well known to all. Most of us know also the common "drone-fly," which so nearly resembles it in size and form, and which we have seen clustering by hundreds on the Michaelmas daisy in the light of a November sun, making the whole air musical with their merry hum, and the very sunlight brighter with their glancing wings. It will be convenient to compare this insect with the bee. Let us, therefore, lay the two side by side. Here (fig. 21, *b*) is the drone-fly with its bright broad wings—it seems no wonder that the creature flies. Now turn to the bee (*a*); four little wisps lie upon its back, and we marvel how it uses them.* The bee is heavier than the fly, the wings may perhaps be of equal, or more than equal expanse when unfolded, but how much force is lost by their division? The drone-fly has one broad wing on each side, the bee has two narrow ones. Why is this? and how is it compensated?

Fig. 21.

a. Drone. Male of Honey-bee.
b. Drone-fly (*Eristalis Tenax*) slightly magnified.

Let us begin at the beginning. When our drone-fly crept from the egg he found himself an uninviting looking little grub with a most inordinate tail, which although it had its uses, by no means improved his appearance. Besides this, he was unfavourably placed, being in the mud at the bottom of a dirty pool, or perhaps of a more dirty drain. Finding it in vain to try to be ornamental, the little grub set about being useful, and began by seeing what could be done with his tail, which, lengthening it with a sort of telescope movement, and elevating it to the surface of the water (he himself being immersed to the depth of perhaps some inches), he found to be an excellent *breathing tube*, through which he might obtain

---

* It was intended that a *worker* hive bee should be figured here, but the figure is of a drone, whose wings are larger. The wings are also partially expanded in the figure, not closed up into a small space as described. The reader must therefore be referred to a live honey-bee when at rest, and to which the following description applies.

a constant supply of atmospheric air, while engaged in his labours below.

Of these labours it must here suffice to say that, feeding on putrifying substances, he was doing his part towards the purification of the world by converting noxious dead matter into the material of an organized, living, wholesome being. At this task he laboured, and if he gained his own advantages thereby, who would grudge them to him, or who could fail to recognise therein the wisdom which has taught each creature to find in his allotted task his allotted share of enjoyment?

Having at length done his work and earned his reward, he next took a little sleep as "pupa," and then burst forth into life, a sun-loving, flower-enjoying, winged creature, with nothing to do but to be happy and nothing to think of but his pleasures. No house building for him—he wants no house. His life will end with the year's warmth and brightness, for he knows no winter. No family cares for him—his children can make their own way in their own muddy pool, be happy and prosperous without his help. Never need he dim his bright wings with sordid labour, or soil his polished body till his little life shall end. And so his broad smooth wings are well suited to his needs.

But now for our bee. He first awoke to consciousness in the form of a fat little, comfortable, lazy white maggot, packed cosily in a waxen cradle ready made for him, with nothing to do for himself or for anybody else, except to open his mouth for the food which careful guardians daily and hourly brought to him, and, deliberately masticating it, to wait for more.

So time rolled on with him, and the one only exertion to which the little sybarite thought of arousing himself

WINGS OF INSECTS, AND THEIR CLASSIFICATION. 49

was that of providing for his own further comfort in the matter of spinning himself a silken nightgown, in which to take his pupa sleep.

Awaking from this the scene was changed, he had become a winged and perfect animal; and into his mind rushed a full sense of his responsibilities. First, he had to make sure of his position in the hive as father, mother, or worker, and finding himself, we will say, a worker, he, or rather she, became aware that the duties of house-builder, housekeeper, nurse, and even of soldier and sentinel, devolved upon her.

The business of life now opened before her, she addressed herself to the task of repaying to futurity that debt which the cares of a former generation had laid upon her, and daily she toiled in its fulfilment.

In these labours not only would the bee have found a pair of wings large enough to sustain her weight a serious encumbrance in some situations, and during some employments, but the wings themselves would have been liable to injury on a thousand occasions, unless the bee had had the power of packing them into a small compass. Therefore, as we have said, she is furnished with

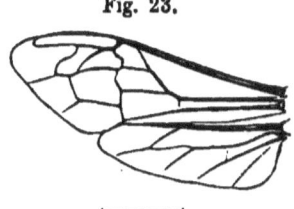

Fig. 22. 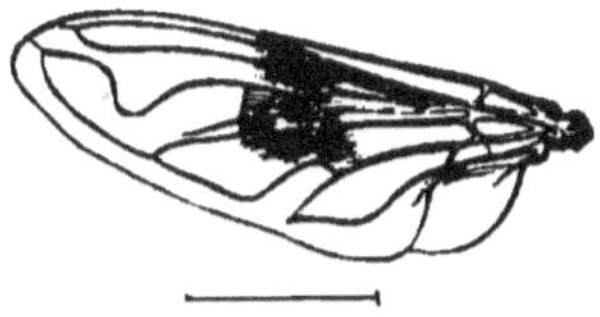
Wing of Drone-fly.
(*Eristalis tenax.*)

Fig. 23.
Two Wings of Bee.

two narrow wings on each side, which lying, when not in use, the one above the other in a small space, pass scatheless "thoro' bush, thoro' briar," as she wends her way in search of food for the day or stores for the future, or as she traverses the narrow passages of the hive, administering food to the ever-ready pupa, or forming fresh cells wherein to lay up her golden treasures.

We see then the purpose of the division of the wing; and now, how is the loss of power compensated? *By the presence of a row of hooks on the front edge of the hind wing, which, fitting into a fold in the hind edge of the fore wing, connect the two in flight and make, as it were, one wing of the two.*

Fig. 24.

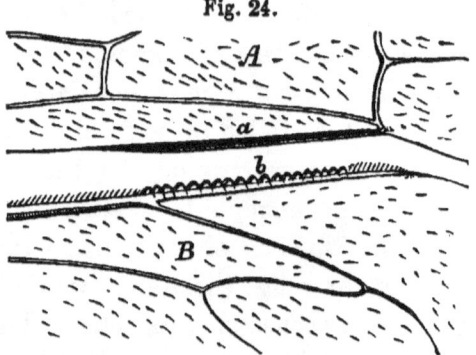

(*A*) portion of hind margin of upper wing of Bee, showing (*a*) thickened ridge for reception of hooks (*b*), on upper side of lower wing (*B*) of ditto.

Here is a beautiful illustration of this variety of structure, with an evident purpose.

For figures of Order VI., Hymenoptera, see Plates VI. to IX.

After the insects with clear and membranous wings, come the tribes of butterflies and moths, and here at least are insects with which all are familiar. Even the painted scales or dust upon their wings cannot pass unobserved, and the Lepidoptera (λεπὶς, a scale; πτερὸν, wing) would seem to require no introduction.

Yet what is the painting upon these beautiful wings, and by what means are the gem-like colours obtained? Let us look into the composition of this common and doubtless very simple little object—a butterfly's wing.

The wing here, as in other tribes, consists of two fine membranes, between which lie the "nervures," a series of tubes, on the nature of which philosophers are hardly yet agreed. Nevertheless, be they veins, or air-tubes, or whatever they may be, it is certain that they perform the mechanical office of bones, in strengthening and supporting the wing.

Attached to the membranes, on both sides of the wings, are innumerable minute scales (the *dust*), each having a little stalk inserted in the membrane of the wing, and all being arranged like tiles, in regular rows, one overlapping another. The variety of form in these scales is very great even in the different parts of each individual; but a distinctive form of scale, generally confined to the male, is found in some genera and species. Some scales are oblong, others triangular, others heart-shaped and tasselled, others in the form of a battledore.

The structure of these scales is next to be observed, and it will be seen that they are not quite so simple as we might have expected, if we believed—as of course we ought to believe—all the instructive little books that talk to us about the "simplicity of nature."

Each of these scales is found to consist of two or three layers of fine membrane. In some the upper layer is more or less covered with granules of colouring matter; in others the second layer is covered with parallel lines, apparently composed of these granules; while in others the second and third layers have the power of reflecting the most brilliant prismatic colouring. By some writers these parallel lines are supposed to be tubes; and let the reader imagine, if he can, the size of a tube of which a large number are found in a single grain of dust on a butterfly's wing! Of these scales, or grains of dust, Leuwenhoeck computed the number on the wing of a single moth to exceed 400,000.

For figures of Order VIII., Lepidoptera, see Plates X., XI.

With the Lepidoptera we enter on a division of insects which differ in one important respect from those which have been already mentioned, and which possess biting jaws. In Lepidoptera (moths and butterflies) the character of the mouth is entirely changed, and these, and all the following orders, may be classed as insects which live by suction. The structure of both kinds of mouth has been described in the second chapter.

The moths and butterflies then, compose the first order of the sucking tribes; the second is that which comprises the cicada, cuckoo spit, aphis, &c.

These have two pairs of wings, which in cicada, aphis, and others, are all alike clear and membranous; while in the froghopper and others the front pair is more or less thickened, somewhat like those of the grasshopper. This is the only order in which the beginner could make a mistake. Attending only to the slight description given here of the wings and wing-cases, he might refer some of the clear-winged insects to Hymenoptera, and others with thickened fore-wings to Orthoptera; but it must never be forgotten that the Hymenopterous and Orthopterous insects are *biters*, whereas these are *sucking* insects, without horny mandibles, and usually provided with a long tubular beak, sufficiently conspicuous to distinguish them with ease. These insects belong to the order Homoptera (ὅμοιος, alike; πτερὸν, wing), so called because, whether the two pairs resemble each other or not, *the wings of the front pair are of the same texture throughout.* By this character they are separated from the order next to be described—Heteroptera (ἕτερος, heteros, different; πτερὸν, wing), in which *part of the fore-wing is thickened, and part membranous.**

For figures of Order IX., Homoptera, see Plate XII.

To the order Heteroptera belong all the rest of those sucking insects which have something like wing-cases, but as has been said, these wing-cases are only partially thickened, and have received the name of *hemelytra*—or half elytra, distinguishing them from the elytra, or perfect wing-cases of the beetles. Of these a good example is found in the common plant bugs, with their variegated colours, or delicate green hue. The water-boatman, the water-scorpion, and many others rank with them.

For figures of Order X., Heteroptera, see Plate XIII.

The next tribe is called Aphaniptera, or not showing its wings (ἀφανὴς, invisible; πτερὸν, wing), which is only a civil way of saying that the insects which belong to it—the *fleas*—have no wings. They have at least none, in the common sense of the word. It has been both suggested and denied that the four wings proper to insects are represented by four scales which grow on the sides of the thorax, and appear to be undeveloped wings. It may be remarked that the want of wings is compensated by the great power of the springing legs, which fully answer all purposes of locomotion.

From the fleas, with their four little scale-like repre-

---

* These two orders were formerly combined, and on their division received their present designations, which accounts for the first being called by a name which distinguishes it from the second half of the old order rather than from the insects of other orders.

sentatives of wings (if they be so), we come to the *flies*, or two-winged insects, Diptera (δὶς, twice ; πτερὸν, wing).

All the insects hitherto described have been shown to possess four wings or their rudiments (unless indeed the fleas be an exception), though these wings are sometimes greatly modified, as in the beetles. The order of Flies, now to be described, clearly possesses but two organs, which can with any propriety be called wings; but even here the deficient wings (which in this case are the hind pair) are supposed to be represented by two little appendages, which grow from the same spot as that which would naturally be occupied by the hind wings, and which are present in no four-winged insect. Of these organs (which are known as the " halteres," " poisers" or " balancers," from one of their supposed uses ; and " malleoli," or little hammers, from their form), little is as yet known. It has been proved, however, that a knot of nerves as considerable as that which supplies each pair of wings in other orders, or the single pair in this, is in connexion with the halteres, and from this it is inferred that they perform some function of importance.

For figures of Order XII., Diptera, see Plates XIV., XV., XVI.

Enough has now been said to show that this small part of so small a creature, even a *fly's wing*, is no simple matter, devoid of interest, or unworthy of study ; and enough to prove that it is but our own ignorance which makes any work of creation small to us—our own blindness which hinders us from seeing the evidences of power and wisdom which lie before our eyes. The above is, however, but a slight sketch of part of the subject; there is much more which might be told of insects' wings.

CHAPTER IV.

THE CHANGES OF INSECTS.

THE changes, or *metamorphoses* of insects, have already been referred to, and it is time to give some account of the various characters of these changes.

In the butterfly and moth tribe they are familiar to all, while perhaps there are many persons to whom the fact will be new, that the process of changing is not confined to butterflies and moths, but that beetles, flies, wasps, grasshoppers, and, indeed, all true insects (see p. 23) undergo changes of like nature, though not in all points the same.

In describing these metamorphoses, the butterfly appears to be the most suitable, because the best known, example.

From the egg of the butterfly proceeds at first a minute caterpillar, which feeds and grows until, having outgrown its skin, this bursts, and the little caterpillar emerges in a newer and larger garment which has been preparing beneath the first. This process (called moulting) is repeated from time to time until the caterpillar has arrived at its full growth; when, by prolonged and apparently distressful exertions, the last skin is burst, and the caterpillar emerges in the form of a chrysalis. The chrysalis lies to all appearance dormant until the time for the animal's final change, when, in its turn, it is cracked from within, and the butterfly comes forth.

All true insects, as has been said, undergo these three changes, but their condition in the imperfect stages is not alike in all, nor will the terms caterpillar and chrysalis always apply to them. We must therefore use the scientific names of *Larva** for the second or Caterpillar stage, and *Pupa†* for the third or Chrysalis stage; the perfect insect is usually termed the Imago.

Both the larvæ and the pupæ differ greatly in habit and in appearance, in the various orders. The larva may be a footless and almost inactive maggot, and even in some cases (as with the social bees, see fig. 25, and ants) be dependent for food upon the care of the parent or nurses. Other larvæ are active and ravenous, and as unlike the perfect insect as possible, as in the case of the water-beetles (fig. 26); whilst in others, as the earwig, grasshopper, cockroach, &c.,

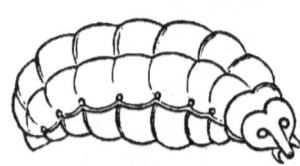

Fig. 25.

Larva of Bee.

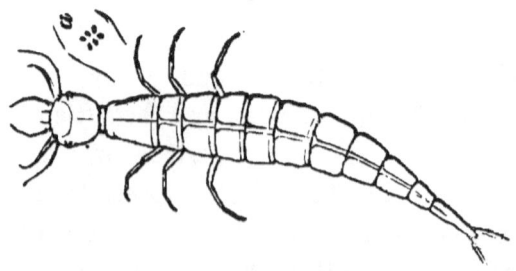

Fig. 26.

Larva of Water-beetle (Dyticus).

---

* From the Latin word *larva*—mask.
† From *pupa*, a child—referring to the swathing-bands of infants.

the larva closely resembles the perfect insect, except in being without wings. Besides these varieties, the larvæ of some terrestrial insects are aquatic in their habits, as in the case of the common gnat, the dragonfly, &c.

The variety of character observable in the pupa is of the more importance, from a scientific point of view, as it coincides (though not without exceptions) with the principal divisions of the insect tribes, and has, indeed, by some of the best writers, been used to mark those divisions. Thus all the beetles (Coleoptera), and all the insects of the bee, wasp, and ant tribes (Hymenoptera) (figs. 27, 28),

Fig. 27.   Fig. 28.   Fig. 29.   Fig. 30.

Pupa of Bee (Front).   Pupa of Bee (Profile).

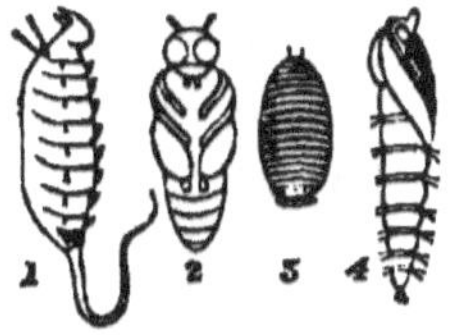

Pupæ of Diptera.
1. Pupa of Drone-fly still in larva skin.
2. Ditto, with larva-skin removed.
3. Ditto of Anthomyia.
4. Ditto of Mycetobia.

Pupa of Sphinx-Moth.

have a pupa which is inactive and non-eating, but which differs from those of the moths and butterflies (Lepidoptera) (fig. 29) and from some of the two-winged flies (Diptera) (fig. 30) in being covered by a skin, which allows the limbs to show separately, as the hand is covered by a glove; whereas in the butterflies and some flies the whole pupa is enclosed in a simple case or envelope. In these, therefore, the pupa in no degree resembles the perfect insect, while in the former it bears some likeness to a dead and wingless specimen.

In other classes of insects the pupa is active, and closely resembles the perfect insect, forming indeed, in most cases, a link between this and the larva. Of this kind are the pupæ of the earwigs (Euplexoptera), and the grasshoppers and cockroaches (Orthoptera), which can only be distinguished from the wingless larva and the winged imago by the rudimental wings, or rather wing-cases, wherein the true wings are being prepared.

To sum up. From the egg of an insect emerges the *larva*, which, whether active and independent, or partially inactive and dependent on others, is always a feeding and a growing animal. To the larva succeeds the *pupa*, which may be totally quiescent and incapable of feeding, or which may be active and voracious, but which never grows or moults. From the pupa proceeds the *imago* or perfect insect, which thenceforth neither grows nor undergoes change or moult.

In the life of some insects the chief part seems to be played whilst in the preliminary states; the imperfect insect preying, building, tailoring, and generally living for a much longer time than the perfect insect; the only business of which, in these cases, seems to be to perpetuate the species and to die; whilst, on the other hand, there are those which, having lived a dependent and inactive life in their earlier stages, take upon them in their maturity all the duties of parents, nurses, governors, citizens, and artizans.

And now, after all that has been written in this and the preceding chapters, it must be confessed that the young student will occasionally find difficulties in his way, even in the first step of determining whether a certain animal be a true insect or not. Again, there are exceptional insects, concerning which he will at first be puzzled to decide whether they are in a perfect or imperfect stage; but a very little experience and observation will do more for him than the addition of many words in this place; and he is advised here, and throughout his studies, to turn at once from the written page to an examination of the objects themselves, and thus to exchange words for knowledge.

TABULAR SUMMARY OF CHAPTERS II., III., IV.

A. Insects with biting jaws.

ORDER I. COLEOPTERA (κολεὸς = *koleos*, *a sheath*; πτερὸν = *pteron, a wing*).
Pupa inactive.
Fore-wings horny or leathery, covering the hind-wings.
Hind-wings with branching nerves, folded lengthwise and across.
Examples. — *Cockchafer, Devil's coachhorse, Ladybird,* &c.

ORDER II. EUPLEXOPTERA (εὖ = *eu, well;* πλεκτὸς = *plectos, folded*).
Pupa active, resembling larva and imago.
Fore-wings leathery, very small, not quite covering the hind-wings.
Hind-wings large, with radiating nerves, folded lengthwise like a fan, then folded twice across.
Example.—Earwig.

ORDER III. ORTHOPTERA (ὀρθὸς = *orthos, straight*).
Pupa active, resembling larva and imago.
Fore-wings parchment-like.
Hind-wings large, with radiating veins, folding like a fan.
Examples.—Cockroaches, Grasshoppers.

[ORDER IV. THYSANOPTERA. See Chapter IX.
Example.—Thrips.]

ORDER V. NEUROPTERA (νεῦρον = neuron, a nerve).
Pupa active or inactive.
Four wings, all clear and membranous, with veins forming a fine network.
Examples.—Dragonfly, Lacefly, &c.

ORDER VI. TRICHOPTERA (τριχιῶν = trichion, hairy).
Pupa inactive.
Four wings covered with hairs.
Hind-wings larger than fore, and folding.
Example.—Caddis-fly.

ORDER VII. HYMENOPTERA (ὑμήν = hymen, a membrane).
Pupa inactive.
Four wings, clear and membranous. Veins branching, not numerous.
Hind-wings smaller than the fore-wings, and connected with them in flight by hooks.
Examples.—Bees, Wasps, &c.

[ORDER VIII. STREPSIPTERA. See end of Chapter VI.
Example.—Stylops.]

B. Insects with sucking mouths.

ORDER IX. LEPIDOPTERA (Λεπὶς = lepis, a scale).
Pupa inactive.
Four wings, large, covered on both sides with fine dust or scales.
Examples.—Moths, Butterflies.

[HEMIPTERA.—Now divided into Homoptera and Heteroptera.]

ORDER X. HOMOPTERA (ὅμοιος = *homoios, alike*).

Pupa active.* Sometimes resembling the perfect insect.

Four wings, all clear and membranous; or the fore-wings slightly thickened throughout.

Fore-wings largest, not overlapping.

Proboscis springing from under the face, near the throat.

Examples.—Aphis, Cuckoo-spit insect, &c.

ORDER XI. HETEROPTERA (ἕτερος = *heteros, different*).

Pupa active, resembling the perfect insect.

Hind-wings clear and membranous. Fore-wings thickened in part, and clear in part, and overlapping each other.†

Proboscis springing from the front of the face.

Examples.—Water-boatman, Plant bugs, &c.

ORDER XII. APHANIPTERA (Ἀφανὴς = *Aphanes, invisible*).

Pupa inactive.

Wingless.

Example.—Flea.

ORDER XIII. DIPTERA (Δὶς = *dis, twice*).

Pupa inactive.‡

Two wings, membranous, clear, and not folded.

A pair of balancers in place of hind-wings.

Examples.—Gnat, Daddylonglegs Housefly, Bluebottle, &c.

---

* Except Aleyrodes.
† There are several exceptions to this rule. The mouth must then decide the order. ‡ Exception, Gnats.

CHAPTER V.

ORDER I.—COLEOPTERA.

THAT great diversity of habit, food, and structure should be found in the order Coleoptera, might be inferred from the fact that the species of Beetles (of which it is composed) greatly exceed in number those of any other order of insects. In England alone there are about 3000 known, and the number constantly increases.

Amongst these then we find inhabitants of the land and of the water, dwellers on the earth and under the earth; we find scavengers and sextons, fierce hunters and sluggish vegetarians, and, strangest of all, we find a servile race content to live in captivity and minister to the needs or luxury of another tribe of animals.

Between the larval and the perfect state of the same species, diversity is also to be found. Thus some, fiercely predaceous in the imperfect, become vegetarian in the perfect state; and the aquatic larva produces a beetle which, though furnished with swimming organs, and certainly most at home in its native element, is yet both able and willing to use the powerful wings with which also it is provided. Even upon land the water-beetles are by no means destitute of the means of progression, though they can hardly be considered graceful, the gait of the water-beetle on land strongly resembling the hurried shuffle, or "scuttling" motion of a frightened turtle.

This variety of habit implies variety of structure, and it follows hence that, with observation, we may learn to recognise those peculiarities of form which attend certain modes of life, and that thus we shall, in some cases at least, be enabled to read, in the form of a hitherto unknown insect, something of its life and character.

The number of British beetles being so great, it is impossible here to enumerate even the families, much less the genera, in any manner which would be instructive or interesting to the reader;* and only a very slight outline of the order will be attempted, illustrated by examples taken from among common beetles, which may be already familiar to the reader, or which he may easily procure and recognise. An examination and comparison of these will enable him to render himself familiar with the characters used in scientific divisions.

The number of the tarsal joints (see p. 35, fig. 9, *e*) is used to divide the beetles into four large sections. These are—

PENTAMERA, in which all the tarsi are five-jointed.

HETEROMERA, in which the four front tarsi are five-jointed, the hind tarsi four-jointed.

TETRAMERA, in which all the tarsi are four-jointed.

TRIMERA, in which all the tarsi are three-jointed.

The word *pseudo* (false) is sometimes prefixed to -tetramera and -trimera, as these are only apparently

---

* In the series to which the present work belongs, one interesting volume is devoted to this branch of entomology. "British Beetles, an Introduction to the Study of our Indigenous Coleoptera," by E. C. Rye.

ORDER I.—COLEOPTERA.

three and four-jointed; a minute and concealed additional joint existing in both cases.

The form and clothing of the joints of the tarsi are to be noted in the examination of beetles.

The antennæ are next to be observed. Some are slender and tapering, or thread-like, (*filiform*, fig. 31, 1); others thickened towards the free end, and club-shaped (*clavate*); or knobbed (*capitate*, 2). In some the last joints are flat and leaf-like (*lamellate*), attached together at one end and opening and closing like a fan (fig. 31, 5); while in others these joints are thick, and much larger on one side than on the other, forming a knob or club with deep fissures (*fissate*, 6).

Fig. 31.

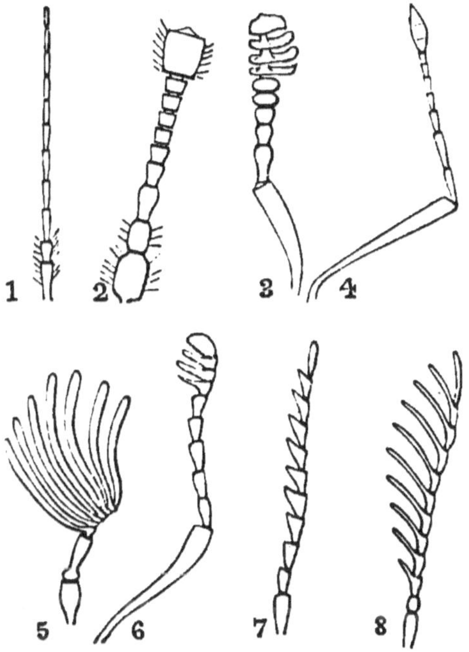

Antennæ.
1. Filiform.
2. Capitate.
3. Perfoliate club.
4. Geniculate.
5. Lamellate.
6. Fissate club.
7. Serrate.
8. Pectinate.

In some the antennæ are slender and toothed more or less deeply, like a saw (*serrate*, 7), or a comb (*pectinate*, 8); and in others the joints present the appearance of a string of beads (*moniliform*). When the knob

F

is formed of thin, flat, distinct joints, as at fig. 31, 3, it is called *perfoliate*, and when the antennæ form an angle, as at 4 and 6, it is *geniculate*, kneed or elbowed.

In beetles of the Section PENTAMERA the antennæ generally afford some indication of the habits of the insects, though exceptions are numerous. Thus, those with slender, filiform antennæ are mostly found to feed on living insects. Those with club-shaped antennæ on dead animal or vegetable matter; those with lamellate and fissate clubs, and also those with slender serrate antennæ, on living plants.

The Pentamerous beetles are divided into four Subsections.*

The first is *ADEPHAGA* ('Αδηφάγος, *adephagos, ravenous*), and contains predaceous beetles, both land and water, which have long horns and two pairs of palpi on their maxillæ. (See fig. 4, p. 65.)

These are again subdivided into land and water beetles.

Of the first Subdivision, *Geodephaga* (Γῆ, *ge, earth*), or the land ravenous beetles, the tiger and violet beetles (Pl. I. figs. 1, 2) are good examples.

The tiger beetle, *Cicindela campestris* (Pl. I. fig. 1, and fig. 3, *a, b*, p. 30), is often to be seen on heaths and sandy roads, and from its great beauty is very unlikely to escape observation. It is easily recognised by its elegant shape and beautiful colouring, and by the remarkable agility of its motions, both running and using its wings with a freedom rare among beetles.

A slender, yet strong-looking little creature, with large eyes, compact thorax, and throat and waist well

---

* See the table at the end of Coleoptera.

marked;—of a glorious green colour, shaded, or rather *illuminated* with crimson and gold, and bearing cream-coloured spots;—long wiry crimson legs with a metallic lustre, and breast and belly clothed with burnished plate armour of bluish green, crimson, and gold:—this is the Tiger-beetle.

The ferocity of this beetle is perhaps as great as that of any animal known. The female has often been seen to deliberately dismember and eat her husband, though it remains a puzzle to naturalists that the husband—an insect apparently equal to herself, or nearly so, in size and power—should submit to this. In captivity the Cicindelæ will (says Mr. Holmes, "Zoologist," 475) "fight savagely, rearing up against one another like dogs. I have known one decapitate his adversary by a single stroke of his jaws." It is not, however, usual for beetles to prey on their own species when alive and not in confinement, though this rule is not without exception. The female may be known by two dusky spots near the base of the elytra, and also by the difference of form in the legs of the two sexes; the tarsi being simple in the female, while in the male the three basal joints are slightly dilated and cushioned.

There are only five British species of the Cicindela, which may be recognised by a pointed claw or hook terminating the maxillæ, and which is found in no other British land beetle (see fig. 4, p. 30). The Cicindela is essentially diurnal in its habits, running and flying freely in the sunshine.

Carabus violaceus (Pl. I. fig. 2), another of the ravenous land beetles, is a large, elegantly formed beetle with a beautiful violet lustre upon the thorax and the wing-cases, which latter, like those of many of the family, are firmly soldered together. It is frequently to be met with in houses haunted by cockroaches and crickets, finding there a plentiful supply of food, the nocturnal habits of these insects (especially the cockroach) agreeing well with its own, as, indeed, with those of most of the Carabi. Predaceous though the Carabus be, it is almost as common to find it half devoured by ants as alive and well, and it appears highly probable that these little creatures attack it when alive, and when one would have supposed it capable of defending itself from their attacks That ants do so attack large living beetles is well known, and the writer once saw a cockchafer under the process of being devoured alive. In this case the whole of the abdomen was gone, and great part of the thorax, only enough being left to hold together the head, wing-cases, and three legs, one on one side and two on the other. With these three legs and this nearly empty *half* of a thorax, the miserable creature was walking about, carrying with him his "detested parasites," which continued their attentions till they were somewhat forcibly brought to a conclusion by the finder.

Another insect of this division is the "Bombardier," which is not uncommon, and attracts attention by a peculiar habit of suddenly ejecting an acrid fluid, as by a little explosion, and which is visible and, at least in the larger foreign species, even audible. The beetle is easily provoked by irritation to these explosions, which however, become weaker when repeated. Mr. Holmes (Zool. 475) mentions the fact that the discharge has been induced so long as four days after death.

The second Subdivision of the predaceous long-horned beetles are the *Hydradephaga* (ὕδωρ, *hydor, water*).

The water-beetles may generally be recognised as

such by their hind legs, which are long, somewhat flattened, tapering, and fringed with hairs; and occasionally present more remarkable modifications of form (see Gyrinus, p. 36, figs. 11, 12), and are obviously fitted for swimming rather than for walking.

The large Dyticus, a beetle common in fresh-water aquaria, is an example of this division, and it, and the smaller and commoner Acilius (Pl. I., fig. 3), much resembling it, and which may be seen floating tail upwards, in almost any pool or duck-pond in the country, are both well-known insects. They are rendered conspicuous by the curious fore-legs of the male, three of the tarsal joints of which are spread out, and together form a nearly circular disc fringed with strong hairs and studded with suckers, forming a singular and beautiful prehensile organ.

In the Dyticus, as in the other predaceous water-beetles, the long oar-like hind-legs are conspicuous and well-marked as natatory organs; and, like most other rapacious animals, the Dyticus is enabled to move very swiftly. This powerful insect, enclosed in plate-armour, swift and ravenous, must be a frightful antagonist to the soft-bodied inhabitants of the waters.

In the Oxford Museum is one which was taken in the act of devouring a young pike longer than itself. A fierce fight between two Dytici is no uncommon sight, and the male frequently falls a victim to the fury of the female, who attacks and eats him. When however this does not take place, the male usually dies first, and is then devoured by his wife.

The larva is a slender, active animal, with a pair of long, sharp, and curved jaws (see fig. 26, p. 57), which make it no less formidable a companion than the perfect insect. A writer in the "Zoologist" gives a rather striking instance of the voracity of one of these insects which, plunged with its prey, a half-dead eft, into strong spirits of wine, continued to eat for twenty minutes or half an hour, during which time he was himself actually dying.—*Zool.* ii. 702.

The merry little companies of the *Whirligig* beetle (Gyrinus natator, Pl. I. fig. 4) can hardly escape the notice of any haunter of shady pools; and the means by which the gyrations of these glittering and silvery globules (as they appear when in motion), are described in a foregoing page (p. 36).

The Gyrinus is small, boat-shaped, and black in colour, and has peculiarities of form besides that in the swimming apparatus. The eyes (fig. 32) are so divided as to give the appearance of a pair on each side of the head — one directed upwards, the other down; a modification which is found in some Dung beetles.

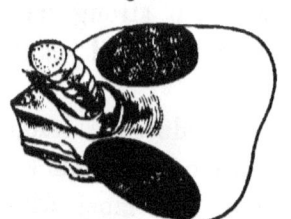

Fig. 32.

Side view of head of Gyrinus.

The antennæ also, are remarkable in form (fig. 33), and the parts of the mouth are well worth examining. The insect is supposed to live on small dead insects, which it seizes when floating on the water

If neither the swiftness of the gyrating motion, nor the beauty of the contrivance which produces it, nor the singularity of the other

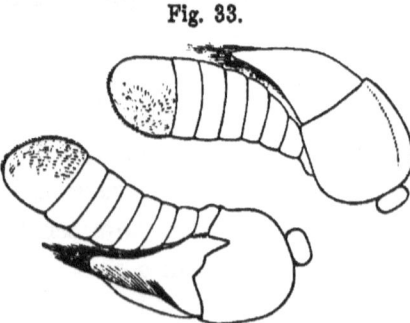

Fig. 33.

Antennæ of Gyrinus in different positions, highly magnified.

parts of the common little Gyrinus natator, serve to impress it upon the recollection of the reader when once seen, he will not easily forget it when once smelt. This remark, however, does not apply to the other species of Gyrinus.*

The eggs are placed end to end in parallel lines upon the leaves of water-plants.

The second Subsection of Pentamera is *RYPOPHAGA* (Ρύπος, *Rypos, filth*), and consists of what may be called scavengers of both land and water. They are distinguished from the preceding (the ravenous beetles) by the form of the antennæ, which are comparatively short and more or less club-shaped, and by the maxillæ, which have but a single palpus.

The first Subdivision of these are the water-lovers, *Philhydrida* (φιλέω, *phileo*, to love; ὕδωρ, *hydor*, water.)

The best known of these is the very large Hydrophilus Piceus or Hydrous Piceus, which greatly exceeds the large Dyticus in size, and is fiercely predaceous in its larval state. The perfect beetle is a quiet and peaceable animal, which, notwithstanding the great strength and completeness of its coat of mail, sometimes falls a victim to its smaller carnivorous brethren.

As in the Dyticus, there is a remarkable enlargement in the fore tarsus of the male Hydrophilus, the last joint forming a large triangular plate, furnished with spines. The second pair of legs is also spinous, as in the Dyticus the tarsi of the second pair correspond with those of the first in being furnished with suckers.

---

* It is supposed to arise from the voluntary emission of a volatile fluid. The same thing occurs in many other beetles, as in the Carabus, the Bombardier, and others.

The larva of this beetle is one of the fiercest hunters of the water.

The Hydrophilus has a habit, singular among beetles, of spinning a flexible silken sac, in which to enclose her eggs (fig. 34). This is described as resembling a turnip upside down; a curved, pointed horn, about one inch long, rising from the upper side of the sac, which, being compressed on two sides, measures about three-quarters of an inch at its widest, and half an inch at its narrowest diameter.

Fig. 34.

Nest of eggs of Hydrous piceus.

There are several other families of non-predaceous beetles both aquatic and semi-aquatic. Some live in the moss and grass by the side of pools and streams, at times freely entering the water and running on the bottom. Others live in or on the muddy bottom; others are found on the stems and leaves and among the roots of water plants. In cases like these the legs are usually adapted for crawling or wading rather than for swimming.

The water-beetles, however, of most marked aquatic form and habit, by no means confine themselves to their own element; those of the carnivorous section especially frequently leave the water at night, but are seldom found on the wing in the daytime.

Of the land beetles with clubbed horns, and which feed chiefly on dead matter of various kinds, the sextons or burying beetles are perhaps the most remarkable, and are certainly amongst the most disagreeable, owing to a disgusting scent which they acquire from their food.

They form the second Subdivision, *Necrophaga* (Νεκρὸς, *necros, dead*; φάγω, *phago, eating*), and are an invaluable portion of the insect tribes, being indefatigable in their labours among putrifying substances, and it needs but little consideration to enable us to realize the vast importance of these labours. Rapidly as the chemical process of decomposition follows upon the death of either animals or vegetables, it seldom in nature outstrips the generation of agents which more than neutralize the attendant evils; agents which in fact turn this decomposition to account under that law of creation which forbids all waste, and exhibits the spectacle of life and enjoyment arising out of death and decay.

To this it is no objection that all the evils which can arise from decay and putrefaction are to be met with in cities and other congregations of men, and that nature provides no means adequate to the demand in such cases. Where men congregate and cause these evils, there human labour is the provided remedy. Nature is no longer the sole agent, art takes her share of responsibility; and, as in the supply of food, so in other works necessary to the wellbeing of man, man depends much upon his own exertions. In nature we find no accumulations of filth, no masses of corruption neglected; we may spend hours—nay, days—in the woods and on the heaths and find nothing to distress the eye; while, even in the overcrowded haunts of men, myriads of little living creatures are at work, giving no inconsiderable help in setting to rights what he has set to wrongs.

The colouring of the genus Necrophorus is likely to attract attention. The elytra are marked with broad alternate bands or patches of orange and black, but the insect is heavily formed, and of unattractive appearance. The Silphæ (another genus of the burying beetles) are very flat insects, of an oval outline and dusky colouring, black ash colour, brown, and grey, being the prevalent hues. (See Pl. I. fig. 5.)

The burying beetle, like the vulture, appears to scent his prey from a distance, and his first endeavour is entirely to bury the carcase of any dead animal left upon the surface of the ground. By scraping the earth from beneath the carcase it succeeds in effecting this, and it is said that the prey is sometimes buried at the depth of nearly twelve inches. When this is done the female deposits her eggs in the carrion, and these, when full-fed, undergo their change into the pupa state whilst still under ground.

It is recorded that in fifty days four beetles buried four frogs, three small birds, two fishes, one mole, two grasshoppers, the entrails of a fish, and two pieces of ox liver.

Some allied genera feed on decaying or dead vegetable matter, and are found in fungi and under the bark of trees, and in cellars on the wine corks; some are even injurious to living plants, and some attack living snails, larvæ, &c.

It is impossible to dismiss this group of insects without notice of the genus *Dermestes*, small beetles of which the bristly larvæ prey upon fur, feathers, woollen cloths, dried flesh, and other such substances—the Bacon-beetle is one of these. There is nothing very remarkable in the appearance of the perfect insect, but the clothing of the larva is one of those marvels of nature which so constantly arrest our attention when least expecting to find subjects of admiration. This little grub—sometimes considerably less than a quarter of an inch in length—is clothed with hairs, various in size and form. Of these some are more or less strong, spinous, and irregularly covered with minuter hairs. This is a very common

ORDER I.—COLEOPTERA. 75

form of insect's hair, and may be seen in the fur of bees, the spines of caterpillars, &c. But the grub in question is also furnished with tufts of hairs of another form. These are clothed with whorls of minuter hairs placed at regular intervals round the shaft. Towards the tip this suddenly enlarges into a somewhat heart-shaped body, above which a number of slender appendages, each attached by a point near their summit and spreading

Fig. 35.

Hair of Dermestes.

out at their base, form a most exquisite pinnacle to the whole. To what purpose is this extraordinary structure?

The ferocious looking "Devil's coachhorse," (Pl. I. fig. 6), is a familiar example of a tribe of beetles distinguished by the shortness of their wing-cases, and which form the third Subdivision, *Brachelytra* (Βραχὺς, *short;* Ἔλυτρον, *wing-sheath*). It is a long slender black creature, something like an enormous black earwig, but without the tail forceps, with a curious habit of turning up its tail, and opening its jaws when in expectation of an attack. Nothing comes amiss to this harpy, which will attack and devour almost any other insect, and has been seen feeding on young toads and other such animals.

A larva, not more than half an inch long, of a species allied to this, was once seen to kill and drag into its hole an earthworm which was eighteen times heavier than itself.

When not predaceous, the beetles of this tribe feed chiefly on dead animal matter, dung, &c.; and some species (as is the case with the preceding carrion beetles) find the high flavour in which they delight in rotten fungi and such substances. In short they are epicures of the *haut goût*, and, were they men instead of beetles, would probably be lovers of Gruyère cheese, high game, and Vermout.

The antennæ are longer than those of most of the Rypophaga, and are either slightly thickened towards the end, or are of much the same thickness throughout.

The larvæ are not unlike the perfect insect; and indeed the perfect insect itself, but for its dark colour, the hardness of its skin, and, above all, the presence of wing-cases, might be mistaken for the larvæ of some other species.

Among these beetles we find an example of insects living in captivity amongst those of a kind altogether different; namely, the Ants, who take captive and hold in captivity several small species of the Brachelytra. It appears that the little guests are treated with great care and attention, and that the only incivility exercised towards them consists, first, in taking forcible possession of their persons, and, next, in frustrating the prisoners' attempts to escape. Both these acts have been observed by Mr. F. Smith (Zool. iii. p. 266), and others.

It is not yet known what is the true relation between the ants and the beetles. It is supposed that the latter exude a fluid useful or pleasant to the ants, and which they suck from them. It is well known that this happens in the case of the Aphides, which are commonly imprisoned by ants in a like manner.

The genus *Claviger* is especially noticeable as a dweller in ants' nests, and, being totally blind, may possibly be well content to live in the home in which it is so well cared for. This genus belongs to a family in which the tarsi are only three-jointed, and the elytra and abdomen are much wider than in the pentamerous beetles with short wing-cases. It has, however, been thought to find its proper place among these.

Another cluster of beetles, with horns more decidedly enlarged at the tip, or club-shaped, constitutes the third Subsection, *Cordylocerta* (κορδύλη, *cordule, a club;* κέρας, *keras, a horn*).

The first Subdivision of these is *Clavicornes* (*clavis, a club, cornu, a horn*), in which the antennæ terminate in a solid or perfoliate knob. This contains the oval-shaped and very convex Pill-beetles (Pl. II. fig. 1, Byrrhus pilula). These are easily known by their rounded form, and by their habit, when alarmed, of drawing their small legs so closely together upon the abdomen as to render them almost invisible. There is a provision for this purpose in the form of the abdomen, which has flattened grooves for the reception of the legs, and in the legs themselves, of which the various joints are grooved to receive each other.

The genus *Hister*, containing some small, squarish, hard, shiny black beetles, sometimes with red or buff markings, sometimes with a metallic lustre, have the same habit of feigning death.

These beetles, though club-horned and feeding on dead animal and vegetable matter, differ greatly from the sextons, not in their form only and the contractile power of their legs, but also in the character of the larva.

The next Subdivision, *Lamellicornes*, comprises the Stagbeetles, the Dung-beetles, and the Chafers.

In the first of these, the Stag-beetle, the three or four final joints of the antennæ are much enlarged on one side, forming a deeply-notched knob or club (fig. 31, 6). The antennæ in these is also *geniculated*, or bent like a knee. The fine stag-like "horns" of this beetle are in reality the mandibles, which are enormously enlarged in the males.

In the dung-beetles and chafers the antennæ are *lamellate* (fig. 31, 5, p. 65), the terminal joints are leaf-like, and lie over one another like the sticks of a fan, having the same power of being spread and contracted. They are not geniculated, as are those of the stag-beetle.

There is a species of dung-beetle (*Geotrupes stercorarius*, Pl. II., fig. 2), so common on heaths, on roads, in fields, and wherever else its peculiar food is to be found, that it can hardly be unknown to the reader. It is hump-backed, slow, and of a bluish-black colour, and is nearly as often to be found kicking on its back and displaying a burnished blue underside profusely garnished with pale-brown parasites, as pursuing its business or its pleasure right side uppermost. In the latter case it may be met crawling slowly along, and occasionally stopping to give one or other leg a sort of weak flourish in the air, like an old gentleman talking to himself, and suiting the action to the word.

Like the sextons, this insect buries the offensive substance which it is its office to render harmless, and in so doing performs the further office of rendering it useful. It forms burrows beneath the masses of dung, carrying into them small pellets in which its eggs are enclosed, and thus separates and spreads the manure in the ground.

The Geotrupes is related, and not very distantly, to the sacred Scarabæus of the Egyptians, and their personification of the sun under the figure of a winged

Scarabæus, bearing a globe upon the head,* is neither more nor less than this animal with the ball of dung which it is its habit to form and roll before it.

The chafers, like the dung-beetles, have leaf-like horns, but differ from them in their habits, the perfect insects feeding on leaves and flowers, while the larvæ also are usually vegetarians; some are, however, to be found among the dung-eaters. The common Cockchafer, or May-bug (Pl. II., fig. 3), a large beetle, with the forepart of the head curved downwards, with brown wing-cases and sides marked with an angular pattern of black and white, is known to every one; and the antennæ of the male are a most beautiful example of the lamellate form. The appearance of white dust scattered over the wing-cases of this insect, and the triangular white patches on the sides of the abdomen, are produced by the growth of snowy white scale-like hairs, thinly distributed over the wing-cases, but lying closely together on the sides.

The cockchafer and dung-beetles are fond of flying late in the evening, but seldom fly by daylight.

The June-bug (*Phyllopertha horticola*) is a pretty little chafer, with green thorax and brown wing-cases. Like the cockchafer, it is extremely abundant, and more conspicuously so, as it flies by day, while the cockchafer prefers the evening. The June-bug feeds upon flowers, especially delighting in roses. The common white Scotch rose, which flowers so abundantly, may sometimes be found with scarcely a blossom which does not contain one, two, or three of these beetles.

The Rose-chafer (*Cetonia aurea*), a large and beauti-

---

* See the vignette at the head of this chapter.

ful shining green beetle, is also to be found in roses (but far less frequently than the Phyllopertha), especially in white and blush roses; and the most fastidious insect-hater could hardly deny that the presence of one of these green gems is a beauty added to the flower. They do not, like the June-bug, devour the petals or injure the appearance of the blossom.

Priocerata, the fourth and last Subsection of pentamerous beetles, has thread-shaped antennæ, generally either tapering or uniform in thickness, and not long. In the males, or in both sexes, these antennæ are commonly serrated, or more deeply toothed like a comb.

These beetles include (with others) the *hard-bodied* Skipjacks, or Elater family, of which the too well-known Wireworm is the larva; and the *soft-bodied* Glowworms, Soldiers, Sailors, and others. Most of the beetles of this section are long and narrow in shape.

The common Skipjack (Pl. II. fig. 4) is a long, slender, hard, uninteresting-looking brown beetle, about half an inch long, with very small legs, and neither throat nor waist, the head being indeed sunk up to the eyes in the thorax. The antennæ are short and slightly serrated.

On the approach of danger, this insect, contracting its limbs and antennæ, falls to the ground, where it lies on its back, motionless and feigning death, sometimes for a considerable time, and, indeed, until it believes the danger to be passed; when, with a sudden click, it springs high into the air, probably alighting on its legs; or, if it fail in this, repeating the spring until it is successful. The point of the breast-plate, which is capable of being slipped in and out of a groove behind it, is the instrument used to effect this leap.

The same power of leaping when lying on the back is

possessed by the aquatic Acilius sulcatus, but this is effected by the mere strength of the insect's spring, and there is no especial provision for it.

The Wire-worm, or larva of the Skipjack, is a long, thin, cylindrical, hard, and eyeless grub, which causes great devastation amongst potatoes and roots of various kinds. The larvæ of some species live under the bark of trees, and in rotten wood. These insects belong to the Subdivision *Macrosterni* (Μακρὸς, *large ;* Στέρνον, *breast*), which also contains some pretty black and red species.

The second Subdivision *Aprosterni* (*a, without,* πρὸ, *pro, in front of ;* Στέρνον, *breast*), consists of soft-bodied beetles with serrate horns. Of these the reddish-yellow "Soldiers," and the red-legged black-elytroned "Sailors" (Pl. III. fig. 1), are, perhaps, the best known, being abundant and conspicuous everywhere. In these insects the head is not concealed within the thorax, the legs and antennæ are longer than in the Skipjacks, and the last joint but one of the tarsus is divided into two lobes. They are nearly as actively predaceous as the pentamerous beetles of the first Subsection, but the maxilla has only one palpus.

Not altogether unlike the dark Telephorus is the male of the Glowworm (Pl. II. fig. 5), a soft-bodied, dusky insect, without however the red legs of the Telephorus, with shorter antennæ, with a head even more concealed than that of the Skipjack beneath the thorax, and possessing the remarkable property of emitting light. The light emitted by the male Glowworm is considerably less than that of the female; but, though this is sometimes disputed, the male certainly does emit light. The writer was once reading by lamplight in a farmhouse in the New Forest when a considerable number of dark, soft-bodied beetles, attracted by the light, flew into the room; some of these were placed under a glass, and while being carried through a dark passage, unexpectedly revealed themselves as Glowworms. The true Glow "*worm*," however, is the female of this beetle, (Pl. Il., fig. 6), and is quite unrecognisable as a beetle to an inexperienced eye. It is a narrow, flat, soft, black insect, about an inch long, and marked down the sides with pale spots; the legs and antennæ are short, the thorax and abdomen not very clearly distinguished from each other, and in the common species there is no appearance whatever of wings or elytra. In fact, the female so closely resembles the larva as easily to be mistaken for it. The larva, however, differs in the form of the legs and the length of the antennæ, and also in being provided with an appendage at the end of the body, which it uses as a foot in walking, like the caterpillars of moths and butterflies. This appendage is peculiar, and is said to be used to cleanse the insect after feeding. It may be observed, even with the naked eye, to leave a minute spot of moisture upon whatever it walks over, not at every step, but at occasional momentary stoppages. The larva and the perfect insect both feed upon snails.

Not only are the perfect male and female Glowworm, luminous, but the larvæ, and, it is said, even the eggs, are so in a slight degree. Dr. Todd, in a paper read before the Royal Society, April, 1824, states that the luminous organ continues to give light for a short period after amputation, and that it is to be re-excited by heat, cold, friction and galvanism; by alcohol, camphor and ammonia. He adds, that when the animal is killed by certain poisons, after all light and life have ceased, another fixed and steady light appears, lasting from 12 hours to 4 days.*

The Devil's Coachhorse and other insects, have at times caused some surprise by appearing luminous; the light, however, in this case arising from their having fed on, or crawled over some dead animal matter in a phosphorescent state, particles of which had remained attached to them. The Glowworm is the only luminous insect known in England, and it is worthy of note that it was capable of exciting Dr. Johnson to write the only poem which is on record as composed by him.

Among the soft-bodied beetles are two common and beautiful little species, which may be found on the blossoms of grass and elsewhere—Malachius æneus and bipustulatus. Both these are remarkable for a series of bright scarlet tubercles which, inconspicuous at other times, swell out from the sides of the thorax and abdomen when the insect is alarmed or irritated, and which have been happily termed "irritation bubbles." The æneus is a dark green oblong beetle, about ⅓ in. long, with a long triangular patch of dull red on the fore part of the elytra. The antennæ of the male are curiously formed, the third joint having a hook-like process, which

Fig. 36.

Antenna of Malachius æneus ♂.

comes down over a projection of the second joint. Malachius bipustulatus is a brighter and sometimes rather brassy green, with a scarlet spot at the tip of each elytron.

---

* *Journal of Science and the Arts,* vol. xvii. 269.

The use of the dilatable bladder-like organs on the sides is not ascertained, but it seems to be generally considered as a means of defence.

These little creatures are eminently predaceous, and of two confined together, only one is likely to be found afterwards if they be left undisturbed for a little while.

There are several small wood-boring beetles which belong to this division. They are generally dull in colour, hard, and somewhat cylindrical in form, and the antennæ vary, being of a thread-like and tapering form, or more or less deeply toothed, or, as in the *Anobium* (Pl. III., fig. 2) (the beetle of which the common " Death Watch" is the larva), approaching to the pectinated club of the Stag-beetle, but without the knee-like joint. The larvæ feed upon every variety of dry vegetable matter, and the round tunnels of the beetle book-worm are but too familiar a sight to his human representative. In the case of books, the devastations of these insects may be prevented by the frequent opening and exposure of the volumes; but it is extremely difficult to stop the progress of the wood-boring species when they have once established themselves within the woodwork of houses, furniture, &c.; and this, too, is not even hindered by the interposition of substances which seem impossible to digest; for, even as the human book-worm finds his way through the heaviest authors, so have these been known to work their way through leaden bullets and the leaden lining of cisterns.

So considerable is the mischief effected by these beetles, that in the choice of woods for shipping, the preference of kinds least subject to their attacks becomes a matter of importance.

Some species attack dried insects, fur, spices, and innumerable other substances.

The production of the ticking sound of the Death Watch is accounted for in different ways. Mr. Westwood considers it to be made both by the larvæ and also by the perfect insects, and, in the latter case, to be a signal between the two sexes. Another author, mentioned by him, attributes it to the larvæ alone, and supposes its purpose to be, to discover how near to the surface of the wood they have bored. That they have some means of ascertaining this, appears from the fact that the burrows usually terminate, and the change of the insects takes place close to the surface of the wood.

CHAPTER VI.

COLEOPTERA.—(*Continued.*)

THE second large Section of beetles is HETEROMERA, subdivided first into *TRACHELIA* (τράχηλος, *trachelos, a neck*), beetles with an apparent throat connecting the head and thorax; secondly, *ATRACHELIA* (α, *without*), beetles with the head sunk up to the eyes in the thorax.

Most of the *TRACHELIA* are showy in colouring and active in their movements; the wing-cases are usually wider than the thorax, and flexible, in this resembling some of the serrate-horned pentamerous beetles, as the Soldiers and Sailors. The antennæ vary, being usually rather long and thread-like, sometimes serrate, or branched, and sometimes inclining to clavate. The last joint of the tarsus is widened and divided into two lobes in many of the Trachelia, while it is always simple in the Atrachelia. The perfect insects are vegetable feeders, and are generally to be found in flowers.

The Cardinal, a handsome red beetle, nearly three-quarters of an inch long, with serrate antennæ, is a common and conspicuous example of this subdivision. It is frequently found on ferns and other plants in May and June. There are two English species, Pyrochroa rubens, which is entirely red above (Pl. III., fig. 8), and P. coccinea, which is red with a black head.

The "Spanish fly," or Cantharis of the Pharmacopœia, is another handsome example of this family, but though occasionally met with in England it is not considered indigenous.

Some families in this division contain insects of unusual form and still more remarkable habits. Ripiphorus paradoxus is a humpbacked, long-legged animal, carrying his shoulders very high and his head very low (notwithstanding the feather-like antennæ with which it is graced), and dressed in a coat much too small for him, the scanty elytra being narrow, pointed, and shorter than the wings, which are left with but slight protection. This beetle is, in its earlier state, a parasite upon wasps, living in their nests, and preying on the young wasp grubs.

Another beetle of unusual appearance is the *Meloe*, or common "oil beetle" (Pl. III., fig. 4). This, though differing much in form from the "Spanish fly," is nearly allied to it, and is said to possess similar medicinal properties. It is a large, heavy, awkward, bluish-black beetle, very common on heaths, and on the flowers in hedgerows. The abdomen has a bloated appearance, and the elytra, which are not above half the length of the abdomen, are convex, and overlapping; the wings are wanting. The antennæ of the males of some species have a distorted appearance. Like the Ripiphorus, this beetle is parasitic; but the eggs are laid, not as is supposed to be the case with that, in the nests of the victim, but under the surface of the earth. When hatched the young larvæ take up their situation on some plant, and availing themselves of the opportunity afforded by the visit of a honey-seeking bee, attach themselves to her body, and are by her transported to her own home to destroy, first the progeny for which she was in the act of collecting food, and then the food itself. This at least is what is now believed to be the case, but the observations made are not as yet perfected, and the history presents some difficulties.

Another beetle of rather remarkable appearance in this division is the Œdemera cærulea, a beautiful greenish blue, or bluish-green beetle, far more elegant in form than the Ripiphorus, but with narrow gaping wing-cases. The thighs of this insect are so thick and swollen as to suggest the idea of great leaping power, which, however, it does not possess. The larva lives under the bark of trees.

In the family Salpingidæ are to be found some beetles, with long snouted heads, much resembling the long-nosed tetramerous beetles; but from these they are to be known by the tarsi.

The *ATRACHELIA*, or neckless beetles, have the elytra of a harder consistence than those last described; they are duller in colouring and less active, sometimes inhabiting flowers, but more frequently dark and damp places, and feeding upon decayed wood, fungi, &c. The antennæ in this division vary, being serrate, clavate, or perfoliate. The "Churchyard Beetle" and the beetle of the mealworm are two common species of this division.

The third Section of beetles is TETRAMERA (or Pseudotetramera), in which the tarsi are apparently composed of only four joints.

The beetles of this Section are nearly all diurnal in their habits.

It is divided into—

1. *RHYNCOPHORA, long-nosed beetles.*
2. *LONGICORNES, long-horned beetles.*
3. *PHYTOPHAGA, plant eaters.*

In the *RHYNCOPHORA* ('Ρύγχος, *rhynchos, beak or nose;* φορεω, *phoreo, to bear*) the head is prolonged (see Pl. III., fig. 51, 5, *a*)—in some species very considerably—into a sort of nose or snout, upon which are placed the antennæ.

The *LONGICORNES* and *PHYTOPHAGA* are without this nasal development. In the former the horns are of great length; and the insects themselves are generally long in proportion to their breadth. The Phytophaga have shorter horns, and are more thickset in figure, being oval, roundish, or somewhat square.

The snouts of the Rhyncophora vary, some being short and flat, while in others, as the common brown nut weevil, the snout alone is nearly as long as all the rest of the body. One of the best known among these is the beautiful and brilliant Diamond beetle of India, and this, although a foreign species, is here spoken of because it is commonly known, and a low magnifying power suffices to show the fine scales of which the prismatic lustre gives so beautiful an appearance to the elytra and thorax. A higher power displays similar scales covering our own little green weevils, amongst the commonest of our beetles, and perhaps the most exquisite insect gems to be found in the country.

This is one of the most destructive groups of beetles. Fruit-trees of all kinds, fir-trees, oaks, hazel, grain, peas and beans, turnips, felled timber, alike are subject to the visitations of various species, and, while the check which they place upon the quantity of fruit-produce conduces, under ordinary circumstances, to the improvement of its quality, it is not rare for whole crops to be destroyed by the labours of the perfect insect and its young.

No part of a plant is secure from the attacks of weevils. One species devours the green and soft parts of the leaves of fruit-trees, another the bark, another the roots. Some destroy the young buds, whether of leaves or flowers, while others gnaw their way into and deposit their eggs within the setting fruit, which is to remain suspended till the time of transformation, when "down will come cradle and baby and all," and the grub, after remaining for a time sheltered in the earth, will return to the daylight in a perfect state. Acorns, nuts, young plums, are easy to find with the little weevil grubs enclosed — while the sheller of peas will bear willing testimony to their attentions to that part of creation. Some roll up leaves, which they have previously nearly severed from the tree, and deposit their eggs therein; others lay them in the ground "convanient" to the roots which are to form the food of the young when hatched. The young of the grain-weevil is found so entirely enclosed within the grain, which bears no mark of its entrance, as to make it seem probable that the egg was laid in the flower and continued to live inside the grain, without, for a time, checking its increase in size.

The pointed snout of the weevil is a powerful instrument, which the little owner well knows how to turn to account. A wood-boring species has been observed in the act of boring a hole in the wood by placing its snout against the part to be bored, and then turning its body round and round, in gimlet fashion, till the work was achieved.

The common pea-weevil, Bruchus granarius, is a small beetle, black above, grey beneath, and with grey legs. The antennæ are straight, and the palpi threadlike. The wing-cases are a little shorter than the abdomen. In Calandra granaria—the corn-weevil—the antennæ are elbowed and the palpi conical. The insect is about an eighth of an inch long, and of a dark-reddish colour. This beetle is terribly mischievous in granaries.

Another, destructive on a grander scale, is Scolytus destructor, a little cylindrical, brown, wood-boring beetle, which attacks elm-trees. In the London parks, in the Champs Elysées of Paris, and indeed wherever these beautiful trees are to be found in their glory, there the ravages of this minute enemy have been conspicuous. For many years a controversy was carried on as to whether the presence of these beetles was the cause or the effect of disease in the trees, but the question seems to have been set at rest in a most satisfactory manner by the investigation and experiments of Captain Cox (see the "Zoologist" for 1858, p. 5995), who, by the simple expedient of spoke-shaving the outer bark down to the mines of the Scolytus, and destroying the larvæ, entirely restored seventeen out of eighteen trees in the Regent's Park assigned to him for experiment by the "Woods and Forests." Many trees in the park had died, and were dying, prematurely, and of the eighteen chosen some were slightly injured, some "very severely," and some "most severely." The greater part of these were recovered by his treatment in the course of five years, in six all were perfectly healthy, with the exception of one which was too far gone for restoration.

There are several other species in this family which attack the bark of various trees. The pine forests of Germany afford an example of what it is in the power of these small creatures to effect. It is recorded that in one year (1783) more than a million and a half of trees were destroyed in the Hartz Forest alone.

The mines of different species have a marked character: those of the Scolytus destructor consist of a tube bored upwards by the mother, of two or three inches long, in which she deposits her eggs. The young larvæ when hatched commence forming horizontal bores on each side of this tube, and at right angles with it. These tubes, usually about sixty or seventy in number, at first close together, spread apart as they proceed, and in them the metamorphosis takes place, the beetles when perfect, not gaining daylight by returning through the tubes, but boring their way out of the tree by a short cut to the surface. The female, having deposited her eggs, is generally found closing the mouth of her burrow with her own dead body. It is remarkable that she never encroaches on the burrow of another individual.

Another species is named Scolytus Typographia, or the Printer, from the resemblance which its mines bear to a printed page.

Several of the weevils have a power of producing a low sound by rapidly vibrating the last segment of the abdomen, and rubbing it against the elytra. Some other beetles are capable of producing sound, as the sexton and the asparagus beetle.

The *LONGICORNES* are generally of a larger size than beetles of the preceding group, and are mostly wood eaters. Among them are some dusky insects of nocturnal habits, but there are several species certain to attract notice by the beauty of their form and colouring. The musk beetle (Cerambyx moschatus) is remarkable both for its size and beauty and for the peculiar scent whence it derives its name. It is of a metallic green, and covered with fine indentations; it is above an inch in length, and somewhat narrow in proportion. This insect is to be found upon willows, where, however, its colour greatly protects it from observation.

Clytus arietis, one of the "wasp beetles" (Pl. III., fig. 6), another conspicuous though smaller insect, is black, with clear bright yellow bands and spots. It is about ½ inch in length, of a rather long oblong form, with long and very active legs.

Strangalia elongata is remarkable for its elegant form. The thorax is long and widest at the base. The elytra are, near the base, much wider than the thorax, and taper towards the end. The back is convex, the legs are long and the colour is a pale yellow, marked with a dusky blackish brown. It is a lively insect, and may be found in flowers (especially those of the umbelliferæ) or on the trunks of the trees whence it has emerged on arriving at perfection. This beetle is nearly two thirds of an inch long. Another species of this group, Molorchus umbellatarum, is interesting as exceptional. The wing-cases are as short as in the Brachelytra, but do not cover the wings, which have not even the usual *shortening fold*, but lie at full length exposed upon the abdomen.

The PHYTOPHAGA (φυτὸν, *phuton, a plant;* φάγω, *phago, to eat*) are less elegant in form than the Longicornes. The abdomen is larger in proportion to the thorax, and the outline of the figure varies from oblong to oval, quadrate, and nearly round. The antennæ are short, and the head is partly buried in the thorax.

Among them are some beautiful species, and one of these is the little Asparagus-beetle (*Crioceris asparagi*), a little oblong beetle, which, in the month of June, when the young plants are beginning to run up into seed, may be found in the asparagus beds by hundreds. The head, horns, and legs of this little creature are black; the thorax is red; a red line runs round the outer edge of the wing-cases, which are black, with three large cream-coloured spots on each, two of which are confluent. When teased, it makes, like some other beetles, a curious creaking sound. The eggs of this species are fixed endways on the leaves, and sometimes one is placed standing end to end on another. They are plentiful enough to do great mischief in asparagus beds. A red species of Crioceris frequents the white lily.

The Cassida viridis, or Tortoise-beetle, is a very pretty little creature, completely concealed under a thin oval shell, slightly concave and broad, which is larger on all sides than the body which it covers. It is of a light but vivid green colour.

The Bloody-nosed-beetle, a common, humpbacked, bluish-black beetle, with broad tarsi, and known by its habit of expelling a drop of red liquid from its mouth, belongs to this division.

The last which shall be mentioned is the Turnip-fly (*Haltica*), or, as it is sometimes called from its habit of leaping, the Turnip-flea, a small active beetle, with large muscular thighs formed for leaping. The larva of this insect mines the leaves of the turnip, and the ravages committed by it are such as very seriously injure the turnip crops. Messrs. Kirby and Spence relate that in 1786 the loss occasioned by them in Devonshire amounted to 100,000*l*. The destruction of a whole crop is a common occurrence, even a second sowing often failing to secure success.

This division contains some semi-aquatic genera, of which the pupæ are aquatic, and the perfect insects live chiefly upon the leaves of water plants, taking flight freely in sunny and warm weather. They may be found below the surface of the water, where they cling to the plants, but they are not furnished with swimming apparatus.

The last and smallest of the principal Sections is TRIMERA (or *Pseudotrimera*), with tarsi composed apparently of three joints only. This section has in the Ladybird a representative as familiar as the common house-fly. It may indeed claim a place among domestic insects, often choosing for its winter quarters the grooves and hollows in the plaister mouldings of our ceilings, which are sometimes filled with clusters, several inches long, of these little beetles.

The ladybird, though usually only common enough to be for its beauty's sake a welcome little visitor, is occasionally to be met with in almost incredible swarms. In the August of 1847 they more or less covered miles of ground in Romney Marsh, and a cloud of them, miles in extent, resembling "a long column of smoke from a steamer," was, from the heights of Ramsgate and Margate, seen hanging over the sea. Next morning the coast was covered with them; five bushels were swept from Margate Pier, and Ramsgate Harbour was in nearly the same state. The next two days found Brighton in the same state. (See the *Times*, Aug. 16, 1847.) Five species were counted in Southend on one of these days.

Similar visitations of ladybirds have occurred at Brighton and in other places on the southern coast in other years, the last being in 1869, when these insects swarmed not only in and about Kent, but were seen in one of the London squares like a cloud passing over the houses. The following extract from the letter of a correspondent of the *Times*, in August, 1869, may be interesting to the reader:—

"During the 14th, 15th, and 16th of this month countless multitudes of the little red beetles appeared upon the coasts of Kent and Sussex. The numbers composing these swarms are utterly inconceivable to those who did not see them. They were most numerous close to the shore—tens of thousands perished in the sea near the land. The beaches, piers, and houses near the shore were covered by the swarms, and in many places the streets and roads looked as if strewn by dark red gravel. This extended far inland, and on Sunday, the 15th, myriads were seen in London and its neighbourhood. But, as I have said, the largest assemblages by far were on the east coast, especially at the points nearest to the Continent. This, be it remembered, occurred at a time when there was a continuous east wind.

"On the Sunday in question a scientific friend of mine, a Fellow of the Royal Society, well qualified to observe and record facts of natural history, was fortunate to witness the actual arrival of one of these swarms. When walking on Dover Pier, after morning service, he observed an enormous multitude of these insects, like a cloud, coming over the sea as if from Calais. They were flying from east to west. Large numbers fell into the water, others covered the pierhead as with a red carpet, but the great mass flew on westward, and, as they passed overhead, looked to those who gazed upwards like the interminable flakes of a thick snowstorm as seen from below. A member of my friend's family had seen a similar occurrence the same morning, when, as she expressed it, the little beetles flew against the east-looking windows of the house like a storm of hail. It would be preposterous to imagine that these swarms of ladybirds had been produced in this country and had flown to sea in the teeth of an east wind, simply to be blown back again!

"When we remember the smallness and feebleness of some of our migratory birds, such as the chiff-chaff and willow-wren, that cross the seas to this country during the stormy weather of early spring, the advent of these swarms of ladybirds is robbed of much of its wonder. But the interesting questions are—Whence came they? Where did they collect in such prodigious numbers? What was the home that fed the larvæ from which the beetles sprang? Or if, as seems probable, they had many homes, what impulse brought together these millions for a common emigration? If you kindly give insertion to this letter, some intelligent observers of nature on the other side of the Channel may perhaps answer some, at least, of my queries."

The services of these little creatures are most considerable. Their larvæ, looking like little black speckled crocodiles, are among the most voracious of insects, and their food is the aphis, which although it has other enemies, seems to be kept in check chiefly by the ladybird itself and by its larvæ. They are peculiarly valuable in hop gardens, hops being very liable to the attacks of these flies. It was interesting to compare the numerous newspaper reports of the "fly" damaging the hop-crop in 1869, with those of the freedom of the crops from fly in 1870, in connexion with the arrival of the Ladybirds in 1869 too late to affect the crops of that year.

It is impossible to find space within the narrow limits

of this volume for much that is interesting relating to the larvæ of beetles, but a few words concerning them are necessary.

They vary in form according to the mode of life laid down for them. Thus, such larvæ as are predaceous, as the terrestrial larvæ of the Carabus (fig. 37), and the aquatic larvæ of the Dyticus (see fig. 26, p. 57), are comparatively light and active in form, and have legs of considerable length and power; while, to go at once to the other extreme, the larvæ of some of the mining and boring species, as the nut-weevil, are footless grubs, merely furnished with tubercles, or small fleshy prominences, which, somewhat like the false legs of the caterpillar, aid the insect in such motion as is necessary.

Fig. 37.
Larva of Carabus.

Others, again, as the underground, root-eating Cockchafer larvæ (fig. 38), are strange, clumsy-looking animals, rendered totally incapable of walking on the surface of the earth by the large, curved, lumpy termination to their bodies.

Fig. 38.
Larva of Melolontha (*Cockchafer*).
(Less than nat. size.)

Some long terrestrial larvæ, as of the Glowworm, the brachelytrous beetles, and such of the Skipjacks as are not subterranean, have their long and slender abdomens supported, like the caterpillar, by a terminal false leg, whilst the Wireworm, an underground larva in the latter family, is hard, stiff, cylindrical, and pointed.

It is not, however, to be supposed that running after food, or crawling after it, or quietly living in its midst, is all of which the beetle larvæ are capable. The Cicindela larva, a strange distorted animal, whose humped shoulders, large head, and great curved jaws form his chief attractions in front, while his hinder parts display another hump ornamented with two sharp hooks (fig. 39), seems to be haunted by some not uncalled-for doubts as to the impression likely to be produced by his appearance, and accordingly conceals himself in a deep burrow, where he awaits such prey as may pass by that way. The burrow, which is frequently found in sunny banks, is cylindrical, and a foot or a foot and a half in depth, and by means partly of his hooks, partly of his legs, he fixes himself at its opening, dragging his prey, when caught, to the bottom.

Fig. 39.

Larva of Cicindela (*Tiger Beetle*). (From Westwood.)

The larva of the Devil's coachhorse digs a deep pitfall in somewhat the same manner, but has not the peculiarities of form so remarkable in the Cicindela. It is a long, flat, slender, many-jointed, six-legged animal, with a large head; altogether greatly resembling the perfect insect, except in the absence of wings and wing-cases, and of any evident separation between the thorax and abdomen.

Whilst the Cicindela is provided with hooks acting like anchors, the larva of the Cassida is furnished at the tail with a long fork, which it is able, when at rest, to turn over, and carry parallel with its back. The use of this appendage would be difficult to guess, had not the insect been repeatedly found with this fork laden with excrement, which, held over the body, forms a screen which completely conceals it.

The species of Crioceris (Asparagus beetle), form this

screen of the same material, and retain it in its place without the help of the fork, and without encrusting their bodies.

Concealment is attained in another way by larvæ in a family allied to the bloody-nosed beetles, which form for themselves a portable tent or case composed of various substances, in this resembling the Caddis-worms, and Clothes moths.

If the habits of the Caddis-worms and Clothes-moths are represented by the larvæ of some beetles, others of the weevil tribe remind us of the gall-making Cynips flies, the knots and lumps so often to be observed in turnips and other roots, and gall-like excrescences upon some leaves being occasioned by them, and serving them as dwelling-places.

The leaf-mining moths also have representatives among beetles. The destruction caused by the Turnip beetle larvæ, arises from their mining the leaves in the early stage, and continuing to do so till the crop is lost.

The great value to man of the labours of some carrion-eating larvæ has already been mentioned. The importance of the aphis-eating Ladybird larva is too evident to be missed; but there are many larvæ commonly considered as mischievous, which, nevertheless, are working assiduously in the interests of man. Thus, the fruit-eating, the root-eating, the tree-killing beetles, are all doing their part towards checking the overcrowding, the overgrowth, and the consequent enfeeblement of the whole vegetable world; and if sometimes a flight of Locusts abroad, or an unusual multitude of Cock-chafers at home, effects a destruction which for the time appears a simple evil, we should do well to remember the Fire of London, and other "unmitigated evils," which we have at length learned to view in their true light.

Before leaving the order Coleoptera, an insect must be mentioned which has much perplexed entomologists—namely, the Stylops. This insect, parasitic in its wingless state in the bodies of bees and wasps, is in appearance, habits, transformations, so peculiar or so little under-

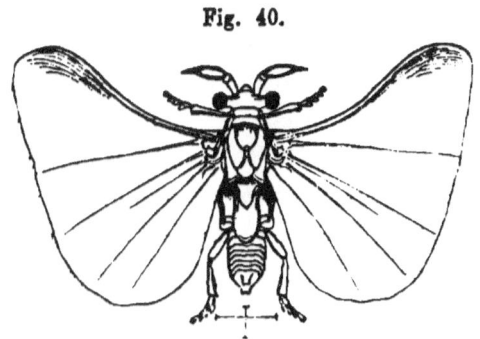

Fig. 40.

Stylops Aterrima, Newport.

stood, that naturalists have had much difficulty in placing it, and it has been moved from one order to another. Mr. Westwood has formed it into an order by itself—*STREPSIPTERA;* but it has more recently been replaced in Coleoptera.

The male Stylops is a singular looking insect, under a quarter of an inch in length, and sometimes very minute; with a pair of enormous *hind*-wings, and no fore-wings, differing in this from the dipterous and all other insects possessing only two wings (as *e.g.*, the exceptional wingless beetles), these having the fore-wings developed while the hind-wings are wanting. That they are the hind and not the fore-wings which are developed, is shown by their position on the thorax relatively to other parts, as the converse appears in the Diptera. As also in the Diptera the missing hind-wings are represented by a pair of hammer-like balancers which grow in their place; so in Stylops, in front of the wings, and situated where the fore-wings would have been, is a pair of curious appendages, supposed to be aborted wing-cases or elytra. These vary in form in different species, and, standing out from the shoulders, add to the singular appearance of the insect. The thorax is disproportionately large, the abdomen small, slender and weak; the antennæ are in various species more or less complicated, being forked or branched; the mouth is very imperfect, if even at all adapted to the reception of food; and the feet are without claws.

The females never acquire wings, and never leave the body of the bee or wasp in which they and the larvæ, whether male or female, are parasitic, swarming sometimes (according to Mr. F. Smith) to the number of 200 or 300. It appears, however, that their presence is not, as in the case of other insect parasites, actually fatal, living bees and wasps being frequently observed with the exuviæ of the perfected Stylops remaining in their bodies, but it is supposed that they destroy the internal organs and render the insects abortive.

The parasite is buried up to its head in the body of the bee, which is usually much swollen, and this head being flattened in shape has something of the appearance of an acarus attached to the bee between the segments of the abdomen.

Fig. 41.

Stylopized Andrena.

When, and how, and where the eggs are laid, is a mystery which remains to be solved.

TABLE OF COLEOPTERA.

SECTION I.—PENTAMERA.—*Tarsi, five-jointed.*

Antennæ long and slender.*
Maxillæ with two palpi.
Habits predaceous.

SUBSECTION I.—ADEPHAGA (Ravenous Beetles).

I. Geodephaga (Land Ravenous Beetles).

Legs formed for running.
  1. Maxillæ ending in a moveable claw.
    Ex.—*Cicindela* (*Tiger Beetle*).
  2. Maxillæ not ending in a moveable claw.
    Ex.—*Brachinus* (*Bombardier*).
      Carabus.

II. Hydradephaga (Water Ravenous Beetles).

Legs formed for swimming.
  1. Front legs short, antennæ long. Fore tarsi of male sometimes forming a disc.
    Ex.—*Dyticus.*
      Acilius.
  2. Front legs long. Antennæ short. Four hind legs greatly dilated.
    Ex.—*Gyrinus* (*Whirligig*).

---

* Except Gyrinus.

SUBSECTION II.—RHYPOPHAGA (Filth-eaters).

Antennæ more or less clavate.
Tarsi of male with basal joints usually dilated.
Habits chiefly scavenger-like.

I. Philhydrida (Water lovers).

Hind legs generally formed for paddling.
Antennæ short and knobbed.
Maxillary palpi long.

II. Necrophaga.

Legs fitted for running.
Antennæ clubbed or knobbed.
Ex.—*Necrophorus (Sexton or Burying Beetles).**
Silpha (Sexton or Burying Beetles).
Dermestes (Bacon Beetle, &c.).

III. Brachelytra.

Legs fitted for running.
Antennæ slightly, if at all, thickened.
Elytra very short.
Body long, narrow, and flexible.
Ex.—*Goerius (Devil's coachhorse).*†
(*Pselaphus* and *Claviger, tarsi three-jointed*).

SUBSECTION III.—CORDYLOCERATA (Club-horns).

Antennæ with large terminal joints.

---

* Elytra rather short and square; club of ant. large, round, perfoliate.
† Ant. obliquely truncated.

I. Clavicornes.

Antennæ ending in a solid or perfoliate knob.
Legs retractile into grooves in the abdomen.
  Ex.—*Byrrhus (Pill Beetle).*
   Hister.†

II. Lamellicornes.

Antennæ ending in a serrate club, or in leaflike joints (Lamellate).
  1. Antennæ elbowed, club serrate.
  Ex.—*Lucanus (Stag Beetle).*
  2. Antennæ straight (Lamellate).
  Ex.—*Geotrupes (Dung Beetle).*‡
   Melolontha (Cockchafer).§ ▪
   Cetonia (Rosechafer).‖

SUBSECTION IV.—PRIOCERATA (Saw-horns).

Antennæ not long, slender, of equal thickness throughout or tapering. Often deeply toothed or comblike.

I. Macrosterni.

Breast-plate long; covering the throat in front; behind drawn out into a point between the legs.
Antennæ short.
Legs short and retractile.
Body hard, head buried to the eyes in thorax.
  Ex.—*Elater (Skipjack).*

---

 * Ant. straight; body oval.
 † Ant. elbowed; body squarish or oblong.
 ‡ Tib. broad and toothed.
 § Tib. slender; claws toothed.
 ‖ Tib. slender; claws simple.

II. Aprosterni.

Breast-plate not covering the throat, nor pointed behind.
Antennæ moderately long, threadlike, serrate, or toothed.
Legs moderately long and slender.
Body usually soft.
> Ex.—*Lampyris* (*Glowworm*).
> *Telephorus* (*Soldiers and Sailors*).
> *Malachius*.
> *Anobium* (*Deathwatch*).

SECTION II.—HETEROMERA.—*Four front tarsi five-jointed; hind tarsi four-jointed.*

SUBSECTION I.—TRACHELIA (with a neck).

Hind part of the head exposed.
Coxæ of forelegs long.
Elytra flexible.
> Ex.—*Pyrochroa* (*Cardinal*).*
> *Ripiphorus*.
> *Meloe* (*Oil Beetle*).

SUBSECTION II.—ATRACHELIA (without a neck).

Hind part of the head concealed.
Coxæ of forelegs short.
Elytra firm.
> Ex.—*Blaps* (*Churchyard Beetle*).

---

* Abd. and Elytr. much broader than thorax.

SECTION III.—TETRAMERA.—*Tarsi four-jointed.*

I. Rhyncophora.

Forepart of head prolonged into a snout.
Antennæ short.
Tarsi cushioned.
Wing-cases sometimes soldered together.
Ex.—*Bruchus (weevil).**
Calandra (weevil).†
Scolytus (weevil).‡

II. Longicornes.

Antennæ long, slender and tapering, simple.
Body long; legs long; jaws large.
Ex.—*Cerambyx*, or *Aromia (Musk Beetle).*§
Clytus (Wasp Beetle).§
Strangalia.∥

III. Phytophaga.

Antennæ short, thread-like or slightly clavate; joints short and distinct.
Head buried to the eyes in thorax.
Ex.—*Crioceris (Asparagus Beetle &c.).*
Cassida (Tortoise Beetles).
Haltica (Turnip-flea).
Timarchia (Bloodynosed Beetle).

---

* Snout short, broad, flat; ant. straight, slender.
† Snout long; ant. elbowed, clubbed.
‡ Snout short; ant. elbowed, knobbed.
§ Eyes kidney-shaped; figure oblong.
∥ Eyes round; figure tapering.

SECTION IV.—TRIMERA.—*Tarsi three-jointed.*

Ex.—*Coccinella* (*Ladybird*).
(*Pselaphus* and *Claviger*, with short wing-cases, placed in *Brachelytra*).

N.B. In this table a few genera only are given to serve as examples.

CHAPTER VII.

ORDER II.—EUPLEXOPTERA.

THE order Euplexoptera contains the Earwigs only; insects as much disliked—and disliked with as little reason (except, indeed, by the horticulturist)—as any of the tribe. The common Earwig is one of the best known of insects, the forceps in its tail affording a means of recognising it at once, at least to those who have nerve sufficient to enable them to look steadily upon it; the less courageous, who sometimes bring stories to the entomologist of encounters with " a dreadful black Earwig, at least two inches long" (if not three), having probably made their observations whilst running away from the Goerius, or Devil's coachhorse, already described.

The Earwigs so nearly resemble the Beetles with short wing-cases, that, except for the tail forceps, they might easily be mistaken for them; indeed, they were formerly classed among Coleopterous insects, an alliance with which seemed pointed out by the cased wings and the character of the mouth. The wings, however, differ greatly in character (see figs. 17 and 18, p. 43); and there is a still more important difference between the Beetles and the Earwigs in the nature of their metamorphoses.

In Coleoptera there is a marked difference between the active larva, the passive pupa, and the winged insect;

whereas, in the present Order, the changes are gradual. In all three stages the insect is active, and the larva, pupa, and imago, have a strong resemblance to each other.

The common Earwig (*Forficula auricularia*), when perfected, is a long, narrow, flat insect, of a brown or puce colour. It has long, slender antennæ of fourteen joints, very short wing-cases, under which are large and beautifully-folded wings (see Plate IV., fig. 1. F. auricularia, with the wings expanded), and, at the tail, a large pair of horny forceps—in the male, strong, dilated at the base, and toothed; in the female, more slender, and quite simple.

The larva, when first hatched, is small, pale-coloured, and active; it increases in size every month, till it reaches the pupa stage. The antennæ are shorter than in the imago, consisting of only eight or nine joints, and the future forceps are nearly straight, long, slender, and feeble. In the pupa the rudiments of wings and wing-cases are apparent, the antennæ are twelve-jointed, and the forceps are strong and curved.

There are four genera of British Earwigs, of which Forficula contains four species, and the others only one each. The genera are chiefly to be distinguished by the number of joints in the antennæ; Forficula having fourteen, Labia twelve,* and Forficesila about twenty-five. The remaining genus, Apterygida, has antennæ of twelve joints, and, as its name denotes, is wingless. The wing-cases, however, are present.

The curious forceps-like appendage of the abdomen

---

* This applies to the single English species, the foreign have from ten to twelve.

seems intended to be generally useful. One correspondent of the "Zoologist" describes the Labia minor, when about to take flight, as turning up its tail, and inserting a point of the forceps under first one wing-case and then the other; by this means quickly unfolding the wings. Another observer, writing to the same journal, reports having seen the common Earwig (*F. auricularia*) seize a small beetle round its middle with the forceps, and carry it away in spite of its struggles.

The reader probably knows that the Earwig is credited with being as careful a mother as the domestic hen; not only sitting on her eggs until they are hatched, but actually covering her young brood like a mother bird. He may not, however, be aware that these facts have been observed and are related by the best authorities, and are not mere popular reports.

Of the other habits of the Earwig it is not easy to speak quite so favourably; the young, for instance, can hardly be said to render due respect in return for such maternal tenderness, as, though professed vegetarians, they have been known to devour the dead body of their mother (Westwood, p. 403). The account of an Earwig carrying off a beetle points also to a carnivorous taste, as it is difficult to imagine any use but one to which his captive could be put.

Flowers are the chief food of the Earwigs, but they by no means confine themselves to this, but consume fruit, and other vegetable productions; indeed, there have been cases when, otherwise, their food must have failed them. There is an account in the "Gentleman's Magazine" for August 1755, of an extraordinary swarm of Earwigs at Stroud:—" There were such quantities of Earwigs in that vicinity, that they destroyed not only the flowers and fruit, but the cabbages, were they ever so large. The houses, especially the old wooden buildings, were swarming with them; the cracks and crevices were surprisingly full; they dropped out in such multitudes that the floors were covered; the linen, of which they are very fond" (!) "was likewise full, as was also the furniture, and it was with caution that people eat their provisions, for the cupboards and safes were plentifully stocked with the disagreeable intruders."

Some doubt has been entertained as to whether the common Earwig ever flies, but it has been found under circumstances which render this probable. It may be that it flies by night, as the lesser Earwig (*Labia minor*) is known to do, these having been observed returning in numbers to their home after the day's work.

The Forficula auricularia and Labia minor are the only British species common. The latter appears to inhabit dunghills and hotbeds. The Forficesila gigantea is a large species which has been found on the sand at Christchurch, but is considered a doubtful native. The apterous earwig is also not common.

Tempore omnia mutantur

CHAPTER VIII.

ORDER III.—ORTHOPTERA.

ORTHOPTERA is the last Order of biting insects in which the hind wings are protected by any kind of wing-case; and the parchment-like and closely-veined *tegmina*, as these are called, seem to form a step between the horny elytra of the Beetle and Earwigs, and the clear and much-veined wings of insects in the succeeding Orders. They differ also in position, the wing-cases in Orthoptera overlapping each other when at rest, while the elytra of Earwigs and Beetles (with a few exceptions) meet in a straight line down the back.

The curious Leaf insects, and Walking-sticks, and the Praying Mantis, are members of the order which have no representatives in this country; and indeed the orthopterous insects known in England are but few, consisting only of the Cockroaches (the "Blackbeetles" of the kitchen), Crickets, Grasshoppers, and Locusts.

In this order (as in the preceding) both larva and

I

pupa are active, and much resemble the perfect insect, the larvæ, however, being without wings, while the pupæ have their rudiments. After the last change of skin the wings and wing-cases are fully developed, except in some species, which in one or in both sexes remain wingless even when arrived at maturity.*

The maxillæ are peculiar in form, having two lobes, of which the upper acts as a kind of sheath to the lower.

The abdomen generally terminates in two bristle-like appendages, short and jointed in the Cockroaches, very long and bristle-like in some Crickets, shorter again in the Locusts.

The English Orthoptera form two groups, the first consisting of the Cockroaches, and distinguished by their *cursorial*, or running legs, which are long, strong, and spinous, and well adapted to this action. The second group consists of the Crickets, Grasshoppers, and Locusts, and is marked by the *saltatorial*, or leaping legs, which are so conspicuous in these insects.

Among the Cockroaches, the common "Blackbeetle," although only too abundant and familiar, is but a naturalized foreigner, and is supposed to have been imported in merchant vessels from the East. Indeed, various other species of these insects are finding their way in the same manner from and into all parts of the world, their omnivorous habits making it easy for them to find subsistence under almost any circumstances. The destruction which they occasion is very great, for even that which escapes being devoured by them they spoil

---

* This occurs in the female of the common Cockroach, which has very short wing-cases, and no wings whatever.

by means of a fluid ejected from the mouth, and which corrodes, discolours, and imparts an offensive smell to whatever has been subjected to its action. The writer has seen the greater part of the contents of a book-case injured in this way, books bound in red or violet-coloured cloth appearing to be especially attractive to the Cockroaches.

The Cockroaches have a curious manner of laying their eggs, not singly, but enclosed in a strong, somewhat bean-shaped capsule, on the outside of which may be seen the impression of the eggs, which lie within in a double row, and in the common Cockroach number about sixteen. The female sometimes runs about for days with this case protruding from her body, a raised serrate ridge along the upper edge of the case helping to retain it in this position. The mother has been observed to assist the young larvæ in making their escape from this capsule.

It would be unfair to suppress any fact which tells in favour of this much abhorred insect, and as there is one yet more abhorred, and unhappily equally domestic in its habits, it may be well to say that a favourite dainty of the Cockroach is the common Bed Bug, and one whose attractions may probably account for its occasional incursions into bedrooms.

Our native species of Cockroaches are much smaller, more delicate, and even attractive-looking insects, in which a careless observer would trace but little likeness to the dark, long-legged, foul-smelling Cockroach of the kitchen. They are found out of doors; some species inland upon herbage of various kinds, others near the sea-shore sheltered under stones, while some are found beneath the bark of trees. B. Lapponia, Pl. IV., fig. 2

(a species to be freely found in the New Forest), is a slender dark insect, with beautifully veined, nearly transparent, pale-brown wing-cases. These are much wider, especially a little below the shoulders, and longer than the body, and give the little creature a very delicate appearance. With the wings closed it is about half an inch in length, but considerably shorter if measured from head to tail with the wings expanded.

The second group of Orthoptera is divided into three families, distinguished by the shape and position of the wing-cases, which either lie flat and horizontally on the back, or shelve downwards, roof-like, at the sides; and by the form and proportion of the antennæ.

Ant. long	Crickets . . .	wing-cases flat.
	Grasshoppers	wing cases shelv-
Ant. short	Locusts	ing.

The most remarkable peculiarity in the wing-cases of this group has been already described (pp. 44, 45)—namely, the musical instrument by which the chirping of these little creatures is produced. The drum-like membrane, or sounding-board of the wing-cases, is however found only in the two first families—Crickets and Grasshoppers; the Locusts, wanting these, produce sound by the friction of their file-like legs against the edge of the wing-cases.

The Crickets form two genera—*Acheta* and *Gryllotalpa*. To Acheta belongs the common House Cricket, (Pl. IV., fig. 3), A. domestica; the larger black Field Cricket, A. campestris; and a small species, A. sylvestris, which is distinguished by the smallness of its wing-cases and the apterous state of the female. In the other genus,

Gryllotalpa, is only one English species, the curious Mole Cricket.

The House Cricket is an active, flattened, long-horned insect, with rather sprawling legs, and the appearance of several tails. These tails consist of, first, the abdominal appendages usual in the order, and which in this are a pair of long tapering bristles; secondly, of the tips of the wings, which being larger than the wing-cases, extend beyond them, when folded, in two long slender points; and thirdly, in the female, of a long ovipositor.

The wing-cases are of a peculiar form in the crickets, being flat along the back and suddenly depressed at the sides for their whole length, thus covering the sides as effectually as the shelving tegmina of the other families.

The bodies of the Crickets are flatter or more depressed than those of the Grasshoppers and Locusts; the tarsi are three-jointed, slender, and spined, so being fitted for running on the ground. In the genus Acheta the ovipositor of the female is long, slender, and projecting; in the Mole Cricket it is withdrawn from sight.

The mole cricket (fig. 42, and fig. 14, p. 37) differs

Fig. 42.

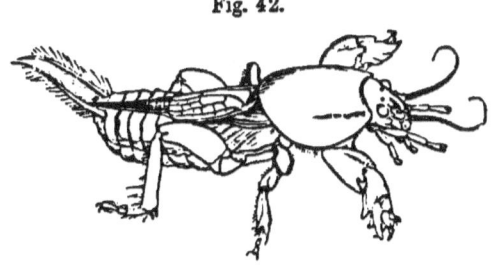

Outline of Mole Cricket.

from the other Crickets most conspicuously in the curious hand-like front legs (described p. 37); which are used by the insect in forming burrows within the earth. Not only do the digging instruments of the Cricket, and its mode of proceeding, resemble those of "the little gentleman in black velvet," but the burrows formed—though not constructed on a precisely similar "ground-plan"—consist, like his, of a neatly finished chamber, approached by winding galleries, and, like the Mole, the Cricket while mining, raises a ridge of earth by which it may sometimes be tracked. Unlike the quadruped, however, the insect is fitted for more than underground life, and though not equal, either in saltatorial or in flying powers to others of its tribe, is able both to leap and to fly, and is possessed of perfect organs of vision.

The chosen home of these curious creatures is the soil in damp fields and gardens (whence their French name of *Courtilière*), or in peat bogs; and their food appears to be various, as they not only feed largely upon vegetables—doing great mischief among the roots of plants, barley, potatoes, &c.—but, like the Grasshoppers, have been known to attack and devour each other. The remains of other insects have been found in their stomachs, and in confinement they have been fed on insects and on raw meat, of which they appear extremely fond.

The female Mole Cricket lays, in summer, an immense number of eggs (according to Resel, 300 or more), which are hatched in about a month. The young remain together underground (during the winter in a dormant state) until all the changes of skin are accomplished, and the wings have attained their full growth, which takes place in the following summer.

The insect is not common in England.

The passages formed by the Mole Cricket are not sufficiently wide to allow of the insect's turning within them. This is compensated by the power of moving backwards and forwards with equal ease, and still more remarkably by the exceeding sensitiveness of the bristles at the end of its body, which act like antennæ, to inform the insect of danger approaching from behind.

Crickets generally have more or less the habit of burrowing, none, however, approaching the Mole Cricket in power, or in architectural skill. The Field Cricket, using its sharp, strong jaws as an instrument, digs a refuge for itself in dry soil, sometimes to the depth of a foot; while the House Cricket excavates passages through the mortar of stone or brick walls.

As might be expected of an insect so domesticated as the Cricket, and so harmless, many superstitions have clustered round it; and if, among the sun-loving Greeks, the Grasshopper was hailed as a friend by gods and men, in our colder clime the Cricket is counted as a fireside companion; and dire are the consequences of murdering one little songster, or of the desertion of our hearth by their numbers. It seems, however, that their music is not at all times, or in all places, equally welcome, as the "Spectator" speaks of the voice of a Cricket as striking more terror to the heart than the roaring of a lion. Probably the roaring of the lion was softened by distance.

The tone of the Field Cricket's song is observed to vary according to the state of the atmosphere; and among the signs of the weather collected by Dr. Darwin, is the sharpness of its sound before rain. This is probably to be accounted for by the action of the damp air contracting, and so tightening, the membrane which forms the drum, or sounding-board.

The two families which remain are the Grasshoppers and the Locusts.

There have been so many changes and interchanges of the names of all these insects (including the Crickets), both in various places and at various times, that the reader will find it necessary to be on his guard when he meets with the various generic names—"Gryllus, Locusta, Acrida, Acheta," &c. Thus *Gryllus*, formerly the generic name of the Crickets, now gives place to *Acheta* as applied to them, and is adopted, under the form Gryllidæ, as the family name of the Grasshoppers, while similar confusing changes have been made with regard to the family Locustidæ. To enter upon these details would be alike tedious and useless while the reader is as yet unacquainted with the animals themselves; and here, as in all cases of the same kind, the first step is to study the animals and familiarize the mind with their distinctive characters. With this knowledge the difficulties occasioned by variety of system and diversity of nomenclature, will become a help rather than a hindrance in the work of obtaining a clear idea of the relations and grouping of animals.

According to Westwood, whose classification is followed here, these insects are grouped (see p. 116) into the families of *Achetidæ*, Crickets; *Gryllidæ*, Grasshoppers; and *Locustidæ*, Locusts; the English names assigned by him not necessarily according with the popular use, according to which most of the Locustidæ, in common with the Grasshoppers, are usually called Grasshoppers.

The Gryllidæ (see Pl. IV., fig. 4,) resemble the

Crickets in having long antennæ, a musical apparatus in the wing-cases of the male, and a projecting ovipositor in the female. The latter, however, differs in form from that of the Crickets, being usually flattened and curved or "sabre-shaped." The wing-case of the female (fig. 43) is simple. The Gryllidæ are more slender in form than the Crickets, and their longer and slighter limbs give them an appearance of greater lightness and activity.

Fig. 43.

Wing-case of Acrida viridissima.

They differ also in their shelving, roof-like wing-cases, in the form of the tarsi, which are broad and fleshy on the under side, and in the number of the tarsal joints, which in this family alone is four.

The English species of Gryllidæ number about twelve, and are found chiefly upon trees, &c. There are among them several in which the wings are either absent or imperfectly developed; and in one wingless species, Ephippiger virescens, the wing-cases are very short indeed, and (a circumstance which renders this insect remarkable) are capable of producing the stridulous sound in both sexes. On the other hand, the pretty little slender green Grasshopper of the oak, furnished with long wing-cases and large wings, is the only species altogether destitute of the musical apparatus in the wing-cases.

The large green Grasshopper is a conspicuous species, measuring $3\frac{1}{2}$ inches from tip to tip of the extended wings.

The remaining family consists of the Locustidæ, or short-horned Grasshoppers, and, in other countries, of the Locusts commonly so called.

The Locustidæ (see Pl. IV., fig. 5) differ from the two preceding families in having short antennæ, no drum and file on the wing-cases of the male, no visible ovipositor in the female. The wing-cases shelve, as in the Gryllidæ; the tarsal joints are three in number, as in the Achetidæ, and the chirping sound is produced by friction of the legs and wing-cases. The English species are found chiefly on the grass.

The *Locusts*, so well known in the history of other countries, are by no means unknown in England, more than one species having found their way here on many occasions. They were seen in Yorkshire in the cold and wet season of 1845. In the year 1846, which was hot, and in 1847, accounts were sent to various papers of their appearance in all parts of the country. In the month of September, 1846, they were found in numerous places in and near London, in nearly every county from Yorkshire to Cornwall, and even in Scotland.

In 1848, again, a flight arrived in the South of England, especially in the neighbourhood of London, and a few Locusts made their appearance in England in the Autumn of 1869.

Happily, however, these visitors, which were of several species, have never yet been known to breed in England, and we may, therefore, refuse to consider them as belonging to our own country.

CHAPTER IX.

ORDER IV.—THYSANOPTERA.

It would be difficult to examine a handful of flowers, whether gathered in the field, the greenhouse, or the garden, without finding a host of minute black insects basking upon their petals, or, sometimes, concealing themselves more coyly in the recesses of the flowers. In either case, however, an examination is sure to end in a tickling sensation first on one part of the face, then on another, and we find that—how we cannot tell—several of the little creatures have found their way from the flowers to our persons.

If a few are shaken from a blossom (a Pink or Carnation is almost certain to contain several), at least one mode of locomotion will soon be observed. Let a single insect be watched, and before long he will probably be observed to form an inverted arch, depressing his body in the middle and elevating his tail. In an instant he is gone, apparently without the wings being called into action; and though he may be found again not very far from the same spot, yet the eye has not followed his movement.

Lest, however, he should be suspected of being destitute of wings, his next proceeding is to stand quite still and begin wriggling his tail in an extraordinary manner, turning it up like a Staphylinus, and from side to side in a manner which no sober minded Staphylinus would think of. The object of this appears to be to assist four slender, fringed and generally veinless wings to flourish themselves in the air in an ostentatious manner, but with apparently little result but that of display, as they seem incapable of motion except with the assistance of the tail; and no tail, however active, could be expected to keep four wings at work in flight.

It is not denied that the insect may fly, but it seems to be doubtful. Observed by the naked eye it might easily be taken for one of the minute beetles with short wing-cases, and the active tail greatly increases the resemblance, these beetles, like the Earwig, using their long and slender abdomens to assist them in the folding and arrangement of their wings.

This little insect is the Thrips (Pl. IV. fig. 6), the gardener's pest, known in greenhouses as the Black-fly, in contradistinction to the Aphis, or Green-fly. The mischief which it effects is considerable both in flowers, fruit, and grain.

Like the Aphis, it sucks the juices of plants, and its attacks are shown in the colourless dead spots to be seen in the petals of flowers, &c.

The place of the Thrips in classification is very difficult to determine. Its transformations are like those of the Orthoptera, the insect being active in all stages and acquiring rudimental wings in the pupa state. The wings differ however from these, and from all others, being, as has been said, generally slender, fringed, and veinless, though the forewings in some species have the appearance of veins and approach Elytra in character. The mouth is a true sucking mouth, somewhat resembling that of the Bugs, Aphides, &c., yet retains enough of the mandibulate character to induce its being placed in the mandibulate section.

The females in some species have a visible curved boring ovipositor; and in some species the male is apterous.

These insects form the order *Thysanoptera* (θυσανοι, *thysanoi, fringe; πτερὸν, pteron, wing*), a description of which was omitted in the earlier part of the work (viz. in chapter iii. and the tabular summary, p. 60), as unnecessary and perplexing to a beginner, from its containing only one small group of anomalous insects.

CHAPTER X.

ORDER V.—NEUROPTERA.

In Neuroptera we come to an order of insects against which no charge can be brought by farmer or gardener, by the owner of orchards, or of timber trees, or of pasture-lands—not one of these can, in our own country, have a word to say against any one of the beautiful tribes contained in the present order.

Abroad so much cannot be said, for to Neuroptera belongs the omnivorous White Ant, so great a scourge to the districts in which it is found; but even in hot climates, this one family is the only considerable exception to the harmlessness, with regard to agriculture, of the insects in this order.

It contains many insects which, some by their beauty, some by their frequent occurrence, have become so generally known, as to have obtained common English names. The Dragonflies, with their netted wings, and slender or flattened, and pointed bodies; the long-tailed May-flies or Troutflies, which may be seen near water on a summer's evening in countless swarms; the speckled Scorpion fly, with its curious pincer-like tail, to be found on every hedge; the delicate green Lacefly, with its tenderly-coloured body, large glistening wings, and glowing eyes, —all these are noted by others than Entomologists.

Among the insects here named, some are to be found

with a history like that of the Beetles, *i.e.*, living an active life in the first stage of their existence, and a quiescent in the second; whilst others, like the Earwigs and Grasshoppers, are active in both these stages, undergoing a less marked metamorphosis.

As a general rule (but one not quite without exception), the imperfect series of changes, *i.e.*, that in which all states are active, is found in such of the neuropterous insects as have their wings either always expanded (as some of the Dragonflies: raised above the body when at rest, in other Dragonflies and Mayflies), or lying flat on the back. This perfect series of changes, *i.e.*, that in which the pupæ are inactive, contains such insects as have the wings, when at rest, *deflexed*—lying over the body like a shelving roof. The Lacefly is an instance of this.*

The Dragonflies are perhaps the most universally known of all these insects, one or other species being nearly always to be met with in the neighbourhood of water, whilst the large size and powerful flight of some, the exquisite form and colouring of others, cannot fail to excite attention. They are even a common object of alarm, and are not unfrequently called " horse-stingers," and believed to be dangerous in their powers of biting and stinging. The truth, however, is that they are all of them (in common with the rest of their order) totally destitute of any instrument with which a sting could be inflicted, and as to biting, one of the largest and most powerful Dragonflies, after long and persevering efforts, and under the constraint and provocation of being held to one spot by force, in order to

---

* *Psocus*, having an active pupa and roof-like wings; and *Panorpa*, having an inactive pupa and wings lying flat, are exceptions.

test his powers, succeeded at last in working his jaws only so far into the skin of the finger which held him, as to produce a slight tinge of blood under the surface.

Under ordinary circumstances it is certain he would not have thought of attempting to bite.

Terrible enough the Dragonflies must be, however, amongst the smaller and feebler tribes of insects. Their larvæ and pupæ are aquatic and exceedingly voracious, feeding on every inhabitant of the water small enough to be attacked. On land, or rather in the air, where, swallow-like, the Dragonfly hunts and seizes its prey upon the wing, they verily are flying dragons; and to a hapless Fly the swift approach of one of these glittering "devil's needles," as they are sometimes called, must be terrible indeed. Their flight is remarkable, the Dragonfly being endowed with the power of changing its forward course, and moving backwards or laterally without the necessity of turning.

There are about fifty species of Dragonflies in England, which are divided into two families.

To the first belong the very large Dragonflies frequently to be met with flying up and down in shady lanes in pursuit of prey, and which measure as much as four inches from tip to tip of their powerful wings, and three or four from end to end of their slender bodies. These are species of the genus *Æshna*, or of the more rare *Anax*. The shorter, flat bodied, dull blue, and golden-brown *Libellula*, with others of the same genus, but of more slender form and brilliant colouring, are also of this family, which is distinguished by the wings being always extended, even when at rest; by the large almost semi-globular head, and the immense eyes which in most cases nearly or quite meet.

The second family may be known by the hammer-like, transversely-placed head, the wide-apart eyes, and above all by the position of the wings, which, when at rest, meet back to back over the back of the insect. The wings differ in form in the two families as well as in position. In the first family (Libellula, &c., see fig. 47, Æshna cyanea) the fore-wings are narrow at the base and wider towards the tip; the hind-wings are broadest at the base, sometimes, especially in the males, forming an angle there. In the second family (Agrion, &c., the smaller species of which have a comparatively feeble flight) the four wings are alike and are very narrow at the base, increasing in width towards the tip. Most Dragonflies have a dark spot or *stigma* on the front margin near the tip of the wing, but this is absent in Calepteryx.

To the second family belong the exquisite little, slender, crimson and sky-blue *Agrions*, most fairy-like insects, which are common everywhere; and the splendid but more rare *Calepteryx*, a more beautiful object than which can hardly be met with in the insect world. The body is about 1¾ in. long, slender, burnished, and of an intense dark steel-blue or dark emerald green, it is almost impossible to tell which, as the glancing light gives one colour or the other to its lustrous surface. The wings are large, clear, and gauzy, with a large dark-brown cloud on each, nearly filling the hinder half, but leaving the base and tip clear. These wings add greatly to the brilliancy of the insect, their delicate and innumerable veins being of the same burnished green or blue as the rest of the body. Add to this the prismatic colours reflected from their membranous part, and the picture is complete. This is the male. The female is similar in form and of nearly equal size. The body is of a brilliant grass-green, burnished like the male. The membrane of her wings is unclouded, and is throughout of a rich golden hue. The beauty of these insects passes description when (as they may be seen in the New Forest) hundreds are on the wing together, darting from side to side of a little rivulet, or reposing in the sunshine.

The difference of colour in the sexes is often very great, as in the broad flat L. depressa, of which the female is golden-brown, and the male dull pale blue.*

The powerful flight of the larger species has already been several times mentioned. An instance has occurred of the capture of a Dragonfly at sea, more than six hundred miles from land, a fact which may give some idea of the travelling powers of this insect, which, even with a favourable wind, must have been severely tried. A Butterfly has been observed following a ship equally distant from the land, and similar facts are on record with regard to other insects; but the case of the Dragonfly is peculiarly interesting, as the nature and habits of its pupa forbid the conjecture that the insect may have been taken on board in this state, and have come to perfection there. It is remarked that the incapacity of Dragonflies to subsist for any considerable time without food, is a proof that the journey must have been quickly accomplished.

They prey, both in their earlier and aquatic, and also in the perfect states, upon other living insects, and are exceedingly fierce and voracious. The unarmed Dragonfly will use small ceremony towards even a wasp,— whilst,

---

* This is owing to a fine powder or *bloom* which covers the male, and which may be rubbed off, leaving his colour the same as that of the female.

however, it must be owned that on occasion the wasp has been known to get the better of his big adversary.

In the larva and pupa there is a special and curious contrivance to enable the insect to seize his prey. This is the lower lip, which is composed of four pieces joined together, and is of great size in proportion to the other part of the mouth. When at rest it is folded up and laid over the mouth, which it entirely covers. When called into action it is unfolded and projected forward, and looks like a large, two-jointed bony tongue, terminated by a pair of very jaw-like plates or nippers toothed at their extremity, and which are supposed to represent the labial palpi. It is with this instrument that the apparently sluggish Dragonfly pupa seizes its living struggling prey, and although the parts common to the mouths of other insects may be traced here, their adaptation in this case is peculiar to the Dragonflies.

The transformation of the Dragonflies is gradual, like that of the Grasshoppers, up to the point when the pupa state is about to be exchanged for the winged, and at this point as sudden a change of nature and appearance is made as in the case of insects with quiescent pupæ, such as the Moths or Beetles. The sluggish mud-coloured pupa ascends the stem of a grass, a rush, or any other stalk or stick of convenient size which rises above the surface of the water. Up this it crawls until it is several inches from the water and conveniently clear of neighbouring plants, or whatever else might interfere with its operations. Here the pupa remains, clinging with its legs to the support, the head upwards and the body hanging down. After a time the skin cracks behind, between or before the wing-cases, and the head and thorax of the enclosed fly are drawn out. Slowly follows the abdomen, bit by bit, and as it emerges, the helpless soft young insect hangs head downwards from the opening (fig. 44), the exposed portion of abdomen lengthening every minute, until it seems certain that the still imperfect fly must drop into the water and be drowned. This, however, is very far from the fact,— no sooner is the insect so far out of the pupa-case that

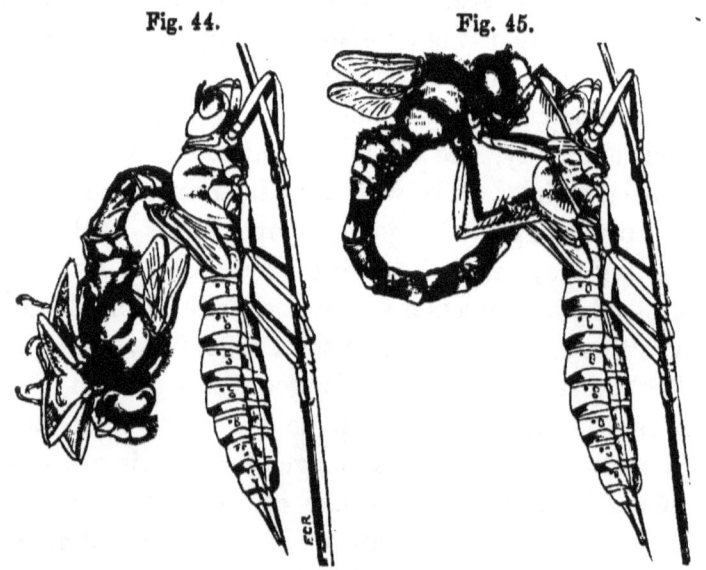

Fig. 44.   Fig. 45.

its fall appears inevitable, than with a sudden effort it curves its body forward and upward, firmly grasps the back of the nearly empty pupa-case (fig. 45), draws the end of its tail out, and stands there, clinging to the now empty pupa-case, which still retains its hold upon the reed (fig. 46.)

It has in this stage a singular appearance. Already the contents of the somewhat broad and flat pupa-case have stretched out into the long and slender abdomen of the perfect Dragonfly. The Dragonfly's head and thorax are there also, differing less in form than the abdomen from those of the pupa—while the wings are four small clouded appendages, little larger than the wing cases of the pupa.

The young fly stands still in the position already described; and as we watch it the wings appear a little and a little larger, until there can be no doubt of their increase in size. Suddenly the insect moves, quitting the pupa-case it walks rapidly up the stalk to which that clings. Whilst this action continues, and for a little while after it ceases, the abdomen appears to become inflated. The fly then becomes quiet, but we observe the inflation of the abdomen to be subsiding and the wings to be so quickly increasing in size that the actual motion is apparent, we *see them growing*. This continues until the abdomen is restored to its former slender shape, when the wings cease to expand; the walk is then repeated, and with the same result, until the four wings have arrived at their full size (fig. 47). The explanation of this proceeding probably is that by the exercise of walking respiration is quickened, and the air vessels in the abdomen are filled with air, which is expelled thence possibly by a voluntary muscular contraction into the wings, and

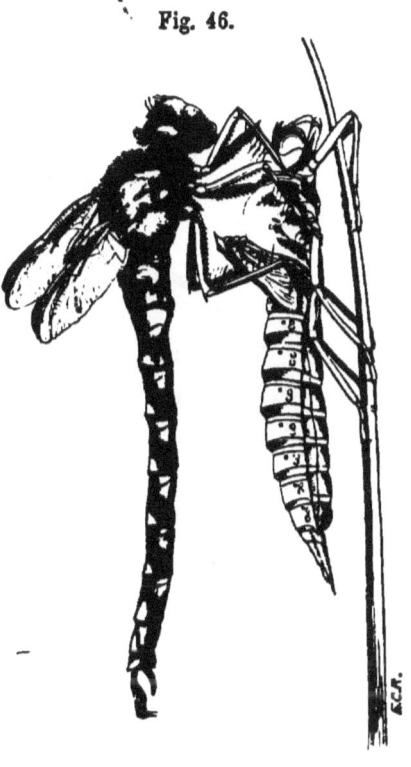

Fig. 46.

by filling the vessels which run through the wing-nerves stretches the wing to its full size.

Fig. 47.

This explanation is partly guess-work, for nothing is known of any internal arrangement by which the air may be pumped into the wings instead of being forced out of the body, but any one who has watched the simultaneous reduction of the apparently inflated abdomen and the apparent inflation, or at least expansion, of the wings, could hardly fail to receive this impression. That blood is forced into the wings during this period is proved by the fact that insects of some kinds will bleed freely through a prick inflicted on the wing whilst it is in the act of expanding, although later the wings may be actually cut or pulled off without any apparent injury to the insect beyond their loss.

While the wings are expanding the Dragonfly assumes a peculiar attitude, so curving its body as to prevent the possibility of any contact between it and the soft delicate young wings.

The process of emerging from the pupa is exceedingly interesting to observe, and it has a curious effect to see the newly excluded insect clinging to what may almost be called its own dead body.

The pupæ of various Dragonflies vary much in form, those of the more or less slender species varying according to the proportions of the imago, though in all cases the pupæ are somewhat shorter and broader than the perfect insect.

The Ephemeron, Mayfly* (Pl. V., fig. 1), or trout-fly is an insect well known to at least all those who live near streams or rivers. Resembling the Dragonfly in the character of its larva and pupa state, it differs from them, and indeed from most Neuropterous insects, in the great inequality of size in the wings, the fore-wings being large and the hind-wings sometimes about one-eighth of their size, sometimes much smaller, or even altogether wanting. When at rest the wings are raised and meet over the back. The Mayflies have slender bodies, short antennæ, and a tail composed of two or three long fine many-jointed bristles.

There is a curious circumstance attending the coming to perfection of the Mayfly. When first emerged from the pupa-case it is quiet and dull—dull in motion and

---

* By Kirby and Spence the Caddis-fly is called "Mayfly."

dim in hue—an almost mud-coloured insect, with large lustreless wings and three shortish tails, may sometimes be seen standing on the wall of a chamber, or in some other situation more or less remote from the water whence it has emerged, and to which it must have flown. In a little while a beautiful insect, with clear and delicately veined wings, is seen standing by the side of something which might be taken for its ghost, so dim and unsubstantial a likeness is it, as with shrivelled and shapeless wings, it stands there in precisely the same attitude, its long tails extended and legs grasping the wall. This *ghost* is a most delicate skin, which enveloped the whole insect, wings, limbs, and all, and enclosed in which the fly had left the pupa-case.

In this way the Ephemeron appears to undergo an extra metamorphosis, but the fact of a delicate membrane covering the insect within the pupa-case is common in other orders, and some insects emerge with more or less of this attached to them. The long-horned bee is a common example of this, as it is usual to find the newly-emerged males with a delicate skin remaining on their antennæ, and which is afterwards stroked off by their spurred legs.

When arrived at perfection the male Ephemera, whose life, as their name denotes (ἐφήμερος, *ephemeros;* diurnal) does not, in some species, extend beyond the day—indeed, seldom beyond a few hours—spends nearly the whole of this brief space upon the wing.*

The mouth is so imperfectly developed that there is

---

* The brevity of the life of these insects was not unobserved by the Ancients; and if the antique gem of which an engraving is placed at the head of the chapter on Lepidoptera (described p. 11), is truly figured, it would seem that the Ephemera, not the Butterfly, is here represented;

reason to suppose them incapable of taking food when in this state. Indeed, there seems little necessity for their so doing, their sole work now being to enjoy the new life on which they have entered; to sport for a few hours in a new element and with new faculties for enjoyment; to perpetuate their species, and to die.

The enormous multitudes of these insects, which sometimes come to life all at the same time, could hardly be believed but by those who have seen them on the wing, literally in clouds, as they may be seen in England. Abroad they are still more plentiful, and Dr. Hagen mentions that on the *Curische-Nehrung* these delicate little creatures are used to feed pigs! Yet of these, says Aristotle, "the least is more noble than the sun, because it hath a sensitive soul in it."

Our knowledge of the Mayfly is at present very imperfect. Not only are there double the usual number of specimens to be studied in each species—*i.e.*, the male and female *sub-imago*, as the first winged state is called, as well as the male and female imago—but all the insects of the family change so greatly after death that preserved specimens are of little value in the study.

The larvæ and pupæ of the Ephemera are of a form somewhat resembling that of the imago (though, of

---

and that the gem signifies, not the escape of the soul from Death, but the shortness of human life.

The engraving is a faithful copy of one given by Guigniaut with the following "explication."

"*Tête de mort* surmontée d'un *papillon*, symbole de l'âme, et ayant à côté d'elle *l'hydrie* qui contient l'eau rafraichissante, conformément aux croyances égyptiennes transplantées en Grèce et communiquées au Christianisme par l'intermédaire des néo-Platoniciens.—CREUZER, Voy. tom. i. liv. iii. p. 403, *et passim;* et tom. iii. liv. ix. Pierre gravée, communiquée à M. Creuzer, par M. Münter, &c."

course, without wings), three beautifully-feathered tails, much shorter than the tail-bristles of the perfect insect, terminate the abdomen, and its sides are fringed with a series of appendages which serve the double office of gills and oars. The pupæ may be known from the larvæ by the wing cases.

In these early stages the Ephemera are predaceous, feeding also, probably, on the decaying animal or vegetable matter which abounds in their haunts at the bottom of ponds and running streams. Some species make burrows in the mud, where they remain on the watch for prey passing by; others are swift swimmers and hunt in the open waters, having in the water, the same faculty as that possessed by the Dragonflies in the air, of altering their course without turning.

The sub-imago has dull wings, fringed with fine hairs, two or three tail-bristles, which are thinly covered with hair, and which, with the legs, are shorter than those of the perfect insect.

The wings of the perfect insect, or imago, are generally spotted and marked with brown, and have a bright surface. In the male the tail bristles and the fore-legs are larger than in the female, the colours are brighter, and the eyes, which are larger, are sometimes so divided as to form two pairs, of which one pair is sometimes considerably elevated above the other. There are three ocelli. Related to the Mayfly is a small family, to which the Genus Perla belongs. Most of these flies resemble the Ephemera in having two tail-like bristles, but they differ greatly from them in the proportion of their wings, the hind-wing in Perla being generally much larger than the fore-wing, and folded when at rest. Besides this the body is less elegant, being rather wide and flat, and of equal width throughout. The males are much smaller than the females, and their wings are small. The larvæ and pupæ of Perla, like those of the Ephemera, are aquatic and active; unlike those, they are carnivorous. The perfect fly is found near palings, and is an inactive, uninteresting looking insect. The "Stone-fly," "Willow-fly," and "Yellow Sally" of the angler are species of the family.

The Laceflies (formerly called Hemerobius, but now divided into several genera), are as conspicuous for their beauty as the Dragonflies. The beauty is however of a very different character. The softness of the parts, the large size and exceedingly delicate texture of the wings, and even the tenderness of the colouring, giving an appearance of great feebleness and fragility to the insect. The one "touch" which lights up the whole is in the glowing eyes, of a golden, sometimes ruby-like lustre, from which is derived the name of one of the genera, Chrysopa, or golden eyes.

The Lacefly (Pl. V., fig. 3) has a cylindrical body, with a small head placed on a neck, long antennæ, slender weak legs, and large broad, lacelike wings, much exceeding the body in length, and, when at rest, lying over it in the form of a sloping roof. The Laceflies are rarely found upon the wing except in the evening, and then may easily be recognised by the cross-like form which they assume in flight, the wings being extended wide and vibrating rapidly, while the progress of the insect is extremely slow and apparently laborious. The species vary in size, the larger measure rather more than $\frac{1}{2}$ in. in the length of the body and about 2 in. from tip to tip of the expanded wings.

The eggs have a singular appearance, being connected

with the leaf on which they are laid by a slender hair-like footstalk about ½ in. long. Six or eight of these are placed near together. The larvæ when hatched feed on Aphides, and it is worthy of note that the Laceflies, like the Aphis-eating Ladybird, have when handled or crushed a strong and disagreeable bug-like smell. The larva of the Laceflies also resembles those of some species of Beetles mentioned above, in the curious habit of clothing itself, using for this the emptied skins of its prey.

As in most of the roof-winged Neuroptera, the pupa state is inactive, and, when about to change, the larva spins itself a silken cocoon from a spinning apparatus which, unlike that of most larvæ, is placed (as in spiders) at the end of the body. The usual position for the spinners of larvæ is at the mouth.

The Laceflies are divided into five genera, containing about fifty species.

Next to the Laceflies comes the only Neuropterous insect which has but little pretension to elegance—namely *Sialis Lutaria* (Pl. V., fig. 4). This insect, resembling the Laceflies in general configuration, is totally without their delicacy of form or colouring. It is of a brown colour, with brownish wings strongly veined; the head is rather large and depressed, and the shoulders are high, giving a very humpbacked aspect to the fly, which is increased by the wings forming a flat surface at the shoulders, from which they shelve into the usual roof-like position. The Sialis is dull and sluggish in its motions as well as in appearance. The larvæ are aquatic and the pupæ inactive.

The beautiful and common Scorpion-fly, *Panorpa communis* (Pl. V., fig. 2), is easily recognised, whether by its long horse-like face, its brown and white speckled, net-like wings, which when at rest lie horizontally over the back, or by the formidable looking scorpion-like pincers which terminate the body of the males. Beautiful to the naked eye, it is still more beautiful when the magnifying of its parts displays the slender legs ringed with even rows of delicate spines, armed with fringed and toothed spurs, and terminated by a pair of curved and comblike claws, somewhat resembling those of certain species of spiders. The head and all the other parts are beautiful, and their transparency, under a very slight degree of preparation, renders them peculiarly accessible to the young microscope student.

Like the greater number of insects remarkable for their beauty, the Panorpa is predaceous. One species at least of the family is said to feed upon leaf-rolling Caterpillars, a kind of prey for the capture of which the toothed claws, and the long pointed head, terminated by a pair of powerful jaws, are well adapted.

The larva and its habits are as yet unknown. The pupa is inactive. The fly itself is found very commonly upon hedges.

There are five English species of Panorpa known. An allied genus, Boreus, contains a curious little insect about the size of a large Aphis, and which, but for the form of its long head, might hardly be recognised as a relation of the Scorpion-fly. It has long legs, and the female is quite wingless, while in the male the wings are reduced to very unwinglike, little curved, leathery, brown appendages. It does not appear to be common.

The Snake-fly, *Raphidia ophiopsis* (Pl. V., fig. 5), represents another family; and though not so commonly observed as most of the insects already named, is as easily recognised when found. About the medium size of a Lace-fly, and with wings somewhat similar but less delicate, it differs from this, and indeed from all others, in the singular length of its neck, which, slender itself, and terminated by a gradually widening and flattened head capable of great freedom of motion, gives a most curious snake-like appearance to the insect. The abdomen is small and short, and the thorax, placed between this and the head and neck, is nearly in the centre of the insect. This snaky look, added to the possession in the females of a long ovipositor, has an uninviting effect; and not long ago the writer received a specimen with an urgent request for an opinion as to the probable extent of the injury which it might have inflicted on a baby on whose face it was found.

The Snake-fly and its larva are insect eaters, the latter living under the bark of trees. The pupa is inactive in its earlier stage, but is said to be capable of walking immediately before arriving at perfection. There are five English species of Raphidia.

The insects hitherto described are probably familiar to the reader in their winged state only; there remains a family of which the larva and pupa, and in some cases a wingless imago, are but too well known. The commonest species of these is a little whitish, semi-transparent creature which we find abounding in books (especially such as are rarely used), old papers, collections of plants, of insects, &c. &c. This little insect is the *Psocus pulsatoria*, the latter name from a sound, similar to that produced by the death-watch, which is heard to proceed from its haunts. There seems to be some doubt, however, as to the fact of this sound being caused by the Psocus. The Book-louse, as the Psocus is commonly called, has always been considered very destructive to the books and collections in which it is found, and although it has found a defender in Dr. Hagen, who (Ent. An. 1861) pronounces the insect, according to his experience, to be nearly harmless, it is difficult to relinquish the suspicion that to its presence may be attributed the destruction of the paste and the brittle condition of the binding in books long unused.

These insects are active in all stages, and the larvæ and pupæ resemble each other, except in the progressively developed wings. Some species, however, never fully develope their wings, the Book-louse being one of these. Others, haunting the crevices of tree-trunks, of palings, walls, books, &c., acquire four large and membranous wings, the expansion of which is sometimes more than half an inch. The females of at least one species are furnished with a spinning apparatus in the mouth, and cover their eggs with a delicate silken web.

These insects are all small, the head large in proportion and triangular, antennæ long, the eyes somewhat large prominent, simple eyes three or none. The body is soft, and generally short and squat; the wings, when fully developed, are large, and have fewer veins than those of most Neuropterous insects.

TABLE OF NEUROPTERA.*

SECTION I.—BIOMORPHOTICA.

Pupa active.
Tarsi with three to five joints.

A. Larva and pupa aquatic.
  * Wings at rest erect.
    a. f. w. large; h. w. small; Tarsi five-jointed.
    Ex.—*Ephemera (Mayflies), &c.*
    b. Wings equal, tapering to base; tarsi three-jointed.
    Ex.—*Agrion, Calepteryx, &c. (small Dragonflies).*
  ** Wings at rest extended horizontally.
    a. Wings nearly equal; h. w. broad at base; tarsi three-jointed.
    Ex.—*Libellula, Æshna, Anax, &c. (Dragonflies).*
  *** Wings at rest, lying flat on the back.
    a. Tarsi, three-jointed.
    Ex.—*Perla (Stonefly), Yellow Sally, Willow fly, &c.*

B. Larva and pupa terrestrial.
  * Wings at rest, roof-like.
    a. f. w. larger than h. w.; tarsi three-jointed.
    Ex.—*Psocus (Booklouse, &c.)†*

---

* This table is borrowed, with some alterations, from a paper by Mr. Newman in the *Zoologist.* Mr. Newman includes in the present Order Phryganea, which, however, in accordance with Westwood's Classification, is here represented as forming the next Order, Trichoptera.

† See page 127, note.

SECTION II.—SUBNECROMORPHOTICA.

Pupa inactive.
Tarsi with five joints.

A. Larva aquatic.
* Wings at rest, roof-like.
    a. Wings nearly equal in size, strongly veined.
    Ex.—*Sialis*.

B. Larva terrestrial.
* Wings at rest, roof-like.
    a. Wings equal, delicately veined.
    Ex.—*Hemerobius* (*Lacefly*).

    b. Wings nearly equal, neck very long.
    Ex.—*Raphidia* (*Snake-fly*).
** Wings at rest, lying flat on the back; equal.
    a. Mouth prolonged into a snout.
    Ex.—*Panorpa* (*Scorpion-fly*).†

---

† See p. 127.

CHAPTER XI.

ORDER VI.—TRICHOPTERA.

No one, whether angler, botanist, or conchologist, can have dabbled much in freshwater rivulets, in pools, or even in ditches when constantly filled with water, without observing certain curious little bundles of dead leaves and sticks, sometimes exceeding an inch in length, and often so apparently shapeless that it is only by the finding of several, all resembling each other, that the attention becomes attracted to them. After a time one of these apparently inert masses begins to move, and a little shiny head and six small legs are seen protruding from one end of the bundle, which is now perceived to be a cylindrical case, irregular indeed externally, but well formed and even within, and terminating at each end in a round opening.

The hermit which inhabits this singular dwelling is the Caddis-worm, well known to anglers, a larva of the family Phryganea, of the order TRICHOPTERA.

The insects in this order have a strong resemblance to certain moths, they have large downy wings which lie close to the body when at rest, greatly expanding when exposed. The antennæ are long, the legs slender, and the colours dull (see Pl. V. fig. 6). A still stronger point of resemblance is in the habit just mentioned of the larva living in a portable case constructed by itself, and carried like the felt tube of the Clothes-moth, or the leafy or lichenous tent of other species of moths. The aquatic habit of the Phryganea larva is alone, however, sufficient to distinguish the insect in this state, while the perfect fly may be distinguished by the *hairy* covering of the wings, the wings of moths being covered with *dust* or minute scales.

The other characters easy to observe, and which distinguish these orders, are the ocelli or simple eyes, which in Lepidoptera are two or none, in Trichoptera three or none, and the organs of the mouth which, though smaller in the case-bearing moths than in other species, are yet fully developed (according to the Lepidoptera form as described in Chapter ii.), while the mouth of the Caddis-fly is almost undeveloped, and is of a totally different type.

These insects vary in size, some being small, others about an inch in length. They run with some activity, but the flight of at least the larger species is uncouth and apparently unenjoying. They frequently enter our rooms at night attracted by the light, and make their presence known by the rustling paper-like sound with which they strike against the ceiling.

Altogether the flies themselves are among the least interesting of insects, unless indeed they acquire interest from the great difficulties which attend their investigation; but as in some other cases the larvæ quite make up any deficiencies of this kind in the perfect insect, and—excepting those of Hymenoptera—the dwellings of the little Caddis-worms are excelled by none in beauty.

They vary much both in material and in the mode of construction. *Phryganea Grandis*, a large and common species, forms an uncouth enough looking case of large, squarish pieces of leaves, arranged however with some regularity in a spiral direction. This case has little beauty when old and brown, and of this the small tailor or tentmaker within seems to be aware. The writer once turned a handful of these creatures, with soddened dingy-brown coats, looking as if made of old tea-leaves, into a glass full of fresh-growing water-plants. It was a most amusing sight to see the eagerness with which the whole party instantly set to work to cut themselves out new coats, which they constructed patch by patch, cutting away a fragment of brown leaf, and sewing on a piece of green leaf alternately, till they all turned out as smart as a party of Robin Hood's merry men. Their appearance during the process, however, was anything but handsome.

Others make their cases of pieces of stick placed either across, sticking out on all sides, or cut into equal lengths, and lying parallel with each other and with the body of the larva, arranged in an exquisite spiral form. Some build up their cases of grains of sand, forming a thin, smooth, shell-like tube, slightly curved and tapering. Others, and these are amongst the most beautiful, cover themselves with small fresh-water shells, and it is really hard to believe that it is unwittingly that they choose the most beautiful forms of these. These cases are all held together, and generally lined with silken threads spun by the laryæ. Some are at liberty in the water, others are attached to plants, &c., and do not move.

A lady has recently made some amusing experiments with the Caddis larvæ. Inducing them to leave their cases by tickling the end of the body (where certain hooks enable the insect to retain their hold of the case); she provided them with fragments of glass, gold beads, and other articles, in which the little creatures soon appeared fully clothed, and doubtless rejoicing in their jewelled bravery.

When about to change into the pupa, the larva sews up the mouth of the case and undergoes the change within it. The pupa, when ready to emerge, acquires so much activity as to gnaw its way out, and rising to the surface it floats to some reed or blade of grass, which it then ascends, and undergoes its final change. The larvæ feed on other living insects and on vegetable matter.

The female has been seen to go a considerable depth under water to deposit her eggs.

There are about two hundred species known in England.

The following table is taken from Westwood's classification.

TABLE OF TRICHOPTERA.

(Family Phryganeidæ.)

A. Antennæ threadlike or pectinated.
   Hind wings not folded.
   *Sub-fam.: Hydroptilides.**

B. Antennæ bristle-like.
   * Hind wings not folded.
   Sub-fam.: Psychomides.†
   ** Hind wings folded.
   a. without transverse nerves.
      1. max. palpi dilated in ♂.
         Sub-fam.: Sericostomides.†
      2. max. palpi, alike in ♂ and ♀.
         Sub-fam.: Ryacophilides‡ (terminal joint ovoid).
         Sub-fam.: Hydropsychides ‡ (terminal joint filiform, very long).
   b. with transverse veins: terminal joint of max. palpi ovoid.

---

* Larvæ enclosed in a flattened, membranous moveable case, with slit-like opening.

† Larvæ enclosed in moveable cases with a circular opening.

‡ Larvæ enclosed in cases fixed to stones.

1. max. palpi very long; very hairy; five-jointed in ♂ and ♀.

*Sub-fam. : Leptocerides.**

2. max. palpi moderate; slightly hairy; four-jointed in ♂.

*Sub-fam. : Phryganeides.**

---

* Larvæ enclosed in moveable cases with a circular opening.

OUT OF THE STRONG CAME FORTH THE SWEET.

CHAPTER XII.

ORDER VII.—HYMENOPTERA.

The insects already described have exhibited great variety in their structure, habits, and instincts; yet, perhaps, all the orders together do not afford more matter for interesting examination than the single order now to be entered upon.

Hymenoptera contains the greater part of those insects which are distinguished by the beautiful modifications of structure which their bodies present; by their social and political institutions; by their domestic virtues, and by their ingenuity as artisans. In other words, of those insects which exhibit the highest development of instinct and of the reasoning powers.

HYMENOPTERA.

The characters of Hymenoptera are as follows:—

The wings are four in number, clear, membranous, and furnished with a few branching veins, but which are sometimes altogether wanting in the smaller species. On the front margin of the fore-wing is a thickened spot or *stigma*, on its inner margin is a fold for the reception of a row of hooks with which the hind-wings are furnished on their front margin, and which, during flight, unite the fore and hind-wings (see fig. 23, p. 49; fig. 24, p. 50).

The veining of the wings in this, as in other orders, is valuable as a help to determining genera, and a figure (taken from Mr. Smith's "Catalogue of British Hymenoptera in the British Museum") will be given in the table of Hymenoptera following Chapter xix. The limits of this work, however, render it impossible even to name more than a few genera, and the characters of the wings can be but very scantily used.

The mouth has been described in the second chapter. The abdomen of the females is furnished with a sawing or boring, or piercing *ovipositor*, or with a venomous *sting*.

This order contains the Bees, Wasps, Ants, Sawflies, Gallflies, and other well-known insects.

In most cases (as in the Bees, Ants, &c.) the larvæ of the Hymenoptera are worm-like grubs, without feet, and live, for the most part, in cells of some kind, formed by the parent for their reception.

These little prisoners are necessarily dependent on the adults for their food, and accordingly we find this provided for them in one way or another. The Solitary Wasps make provision beforehand, by storing up coils of half-killed caterpillars in the clay-built tubes which shelter their young. The Social Bees, Wasps, and Ants, day by day, or rather hour by hour, supply their young with food ; while, in the case of the Gallflies, it is produced by a mode of oviposition which secures for the young both a home and a supply of food, by an arrangement as cosy as that of a little mouse in a big cheese.

The larvæ of one family, the Sawflies, differ greatly from all others in the order. They have both the true and the false legs which are found in the caterpillars of Moths and Butterflies, and, provided with the means of locomotion, seek from leaf to leaf and from branch to branch that nourishment which is rendered scarce only by their own devastations.

The pupa in this order is inactive, and resembles that of the Beetles, the limbs being all sheathed separately (see figs. 27, 28, p. 57), and not as in the chrysalis of the Moth or Butterfly, inclosed in one general or undivided envelope.

The Order is divided into two Sections, TEREBRANTIA, and ACULEATA.

Section 1. TEREBRANTIA (from *Terebra,* an *auger* or *piercer*), consists of insects the females of which are furnished with an ovipositor in the form of a saw, an auger, or other boring instrument. These are the Sawflies, Woodborers, Gallflies, and Ichneumons.

Section 2. ACULEATA (from *aculeus,* a sting or prickle), consists of insects of which the females and the neuters (*i.e.,* the imperfect females) are furnished with a *sting.*

These are the Ants, Sandwasps, Wasps, and Bees.

CHAPTER XIII.

HYMENOPTERA.—TEREBRANTIA.

THE first section of Hymenoptera, TEREBRANTIA, is divided into two subsections, named from the food of the larvæ, *PHYTOPHAGA (φυτὸν, *phyton, a plant ;* φάγω, *phago, to eat*), and ENTOMOPHAGA (ἔντομος, *entomos, an insect*). PHYTOPHAGA consists of the Sawflies and the Woodborers ; ENTOMOPHAGA contains the gall-making insects,† the Ichneumons (parasites which lay their eggs in the bodies of other living insects), and the Ruby-tails.

The insects belonging to these two subsections can at once be distinguished from each other by *their waists being large or small.* That is to say, while in the Sawflies and Woodborers (*PHYTOPHAGA*) the abdomen is attached to the thorax by its whole width (see Pl. VI., figs. 1, 2), in the Ichneumons, Gallflies, &c. (see Pl. I., figs. 3 to 6), the thorax and the abdomen are connected by a small point of attachment, or sometimes by a longer or shorter stalk.

It may be also observed here that no large-waisted

---

* The perfect insects in Terebrantia feed chiefly or entirely on vegetable matter, as honey, pollen, juice of fruit, &c. In some cases they take little or no food.

† The actually gall-making insects are not entomophagous, the larvæ feeding on the vegetable matter of the gall, but the gall-making species are so closely connected with parasitic species that it has been found inconvenient or impossible to separate them. They are therefore included under the head Entomophaga.

insects, resembling the Borers and Sawflies, are found in ACULEATA, the second great section of Hymenoptera.

Subsection I.—PHYTOPHAGA.—The Phytophaga are divided into the Sawflies, and the Borers, otherwise called Leaf-eaters and Wood-eaters. These are distinguished from each other by the ovipositor, which in the Sawflies is in the form of a pair of fine saws, while in the Wood-borers it is a sort of auger. They are also to be easily distinguished by the form of the lip (labium), which is trifid in the Sawflies, simple in the Wood-borers. The tibia of the fore-leg in the Sawflies has two spurs, in the Wood-borers one. These insects vary also in the form and proportion of the thorax.

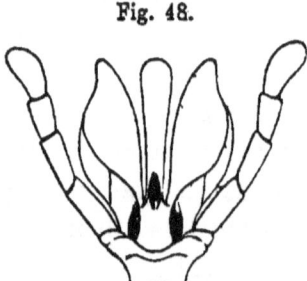

Fig. 48.

Labium of Sawfly—*Trichiosoma.*

Division I.—THE SAWFLIES.—The Sawfly, or Tenthredo* (Pl.VI., fig. 1) is amongst the most easily recognised of insects, its form and general appearance readily distinguishing it from all other insects except its allies the Woodborers. From these, as has been said above, it may at once be known by the two spurs on the fore leg, a character the more useful as it does not entail the necessity of dissection, and is available in either sex.

The body of the Sawfly is of nearly equal width throughout, the head usually, but not always, rather narrower than the thorax. The thorax and abdomen are nearly equal in width, and the sides of the abdomen

---

* The name Tenthredo is used because it is familiar as having formerly been that of nearly the whole family. The name is now restricted to one typical genus of the Tenthredinidæ.

in many instances nearly parallel, except towards the end, which is always pointed.

The antennæ vary considerably, not only in the genera, but even in the sexes. They are sometimes club-shaped (Cimbex), sometimes long and thread-like (Tenthredo) with many joints of nearly equal length; whilst sometimes (Hylotoma), they consist of but three joints, two very short, and the third forming nearly the whole of the antennæ. The number of joints varies from 3 to 30. The antennæ of the males are sometimes exceedingly beautiful in form. In one species it is feathered and resembles that of the Silkworm-moth. In another, the last joint (which forms nearly the whole length of the antennæ) is forked from its base, giving the insect the appearance of a pair of double straight horns. The wings are large, and when in repose lie horizontally on the back, overlapping each

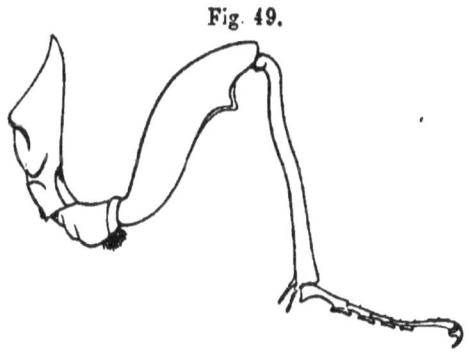

Fig. 49.

Leg of Sawfly (*Trichiosoma*).

other; the legs are of moderate length, or rather long, and have a series of curious sucker-like appendages attached to the tarsus. In some, if not all the species, the claws are cleft or forked.

In colour and markings the species vary much. *Tenthredo viridis, scalaris,* and *punctatus* are of a beautiful green, the two latter with delicate black markings; *Athalia rosæ* has the head and thorax black, the abdomen yellow, whilst *Dosytheus eglantinus* is alternately black and orange. The ground colour most common is black, with yellow wasp-like rings (as many of the *Allantus*, and *Tenthredo zonata* fig. 1, plate VI.), or broader bands of red, as *T. Scutellatus*, which has also pale yellow spots on the thorax. The legs and antennæ are sometimes beautifully variegated with red or yellow and black. The male and female sometimes vary in colour, as in *T. lividus*, of which the female is black, with one pale spot on each side of the base of the abdomen, and a white band towards the end of the antennæ; the male black and red, with red antennæ.

The Sawflies are a large family, the larvæ of which are only too well known both by gardeners and by farmers. The armies of " caterpillars " which in a few weeks or even days, will strip every leaf from a plantation of gooseberry bushes, or rather, which will strip the green soft part from every leaf, leaving the leaf-ribs standing bare on their stalks, are the larvæ of a Sawfly, *Nematus grossulariæ*.

The "turnip-fly,"* *Athalia spinarum*, an insect with black and yellow thorax, black tipped, yellow abdomen, black head and antennæ, and yellow legs, is another of the Tenthredo family, and its devastations are sometimes so great that, as is mentioned by Mr. Westwood, an instance has been known in which many thousand acres of land were obliged to be ploughed up. A very vivid picture of the appearance of a swarm of this species in the winged state is given in a note which Mr. F. Smith,

---

* The name "Turnip-fly" is perhaps more commonly applied to a little hopping Beetle (*Haltica nemorum*), which is very destructive to the turnips.

of the British Museum, has kindly sent to the writer, with permission to use it in these pages.

"*Flights of Insects observed on the Sand-hills near Deal.*—There is, perhaps, no locality with which I am acquainted, more productive of Entomological phenomena than the range of Sand-hills that lie between Deal and Sandwich. It is also one of the richest in the number of species of Coleoptera as well as of Hymenoptera.

"On these hills towards the end of Autumn, clouds of winged ants may be seen, clouds such as I have never witnessed elsewhere. Such, on one occasion, was the case, about the middle of September last, when myriads of Formicida and Myrmecida filled the air. People were fairly driven off the hills by the multitudinous host. The wind on this occasion was little more than an occasional gentle breath from the south.

"On the turning of the tide a line of Ants was left along the shore at high water mark, which I traced to the extent of two miles. I have no doubt it extended all the way to Shellness, which lies full four miles from Deal. On that morning millions of Ants must have perished on the downs. Occasional assemblages of Coccinellida in multitudinous hosts are to be seen along the shore;— similar numbers of Curculionedæ also occur, the species consisting principally of Silona lineata, S. tibialis, S. hispidula, with a liberal sprinkling of Hypera variabilis, &c. Remarkable as all the above mentioned assemblages certainly are, all that I have previously witnessed was eclipsed one morning towards the end of August last, by the sudden appearance of clouds of the Common Turnip fly, Athalia spinarum. I had walked down to the Sandhills, for the purpose of bathing, about 10 o'clock in the morning. On approaching the hills, I was first struck by observing the asphalt pathway that runs along the top of the bank of shingle, being thickly strewn with specimens of Athalia, and on nearing the hills I observed that they were partially obscured by a dense cloud which, shifting in the sunlight, occasionally assumed a bright orange tint, then quickly became of a bright glittering silvery hue, as the sun gleamed upon the shining wings of the hosts of Athalia. I pursued my way penetrating into the cloud of insects which, when observed from a position in which I faced the sun, assumed a tint approaching vermilion red. The Insect-clouds were borne seaward by a gentle south land breeze. I plunged into the water, and hoped by swimming from the shore to free myself from their annoyance, but finding that at a distance of not less than three hundred yards the surface of the sea was thickly covered with them, and as far as I could see that they floated in equal numbers, I hastened to shore and as quickly as I could made my way to the west of the hills, where I found myself freed from their annoyance. Every blade of grass, every rush and twig, was thickly studded with the flies and was bending with their accumulated numbers. The majority of the insects I observed were females. I regret that I did not at the time examine the insects more minutely in order to ascertain whether the flies had deposited their eggs previous to their being borne on the wind to perish in the sea."

The ravages of these insects are not confined to the plants already mentioned; others of the cabbage tribe, with rose trees, willows, apple, pear, and cherry trees, the white thorn, the alder, beech, birch, pine, elm, and aspen, with others both in England and abroad, are infested by species of the Sawfly, and the larvæ, though most commonly feeding on the leaves or stems of plants, as, for example, on the stalks of wheat, have been found inside young fruits, as by Réaumur in pears and by Mr. Westwood in apples.

And now, what do these little creatures contribute towards the justification of the boast with which this chapter commences? They are distinguished by no remarkable display of instinct, nor, at least in England,* by much variety of habit, and the architecture of such species as construct any kind of nest at all is of a very simple character. It is then to their structure that we turn in our search for some matter of especial interest.

Here we find, in the instrument from which the Sawfly derives its name, one of the most beautiful of all the contrivances that have been observed for the placing of the eggs of insects—an instrument from which (if the chronology of Arts and Sciences would allow us to believe that optics had ever been in advance of mechanics,) we might suppose that man had borrowed not the idea only, but the perfect pattern of the *Saw*.

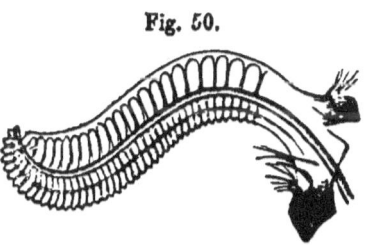

Fig. 50.

Single Blade of Saw of *Cimbex*.

---

* There is in the "Zoologist" (609) a curious account, by John Curtis, Esq., of the proceedings of the gregarious larvæ of a Tenthredo in Brazil, and which scarcely yields in interest to the well-known histories of the concerted architectural labours of Bees, Ants, &c. There is one remarkable variation, however, in this case—namely, that while the chambered palaces of the Bees, Wasps, and Ants are nurseries built by the perfect insect for the rearing of the young, the edifice of the Tenthredo is less a palace than a *tomb*, being built *by the larvæ* for their reception during the sleep which precedes the last metamorphosis.

If so charming a writer as Mr. Kirby (Spence?) could do no better than borrow a description of this wonderful little tool, it is hoped that a humbler writer may well be forgiven for following in his steps, and for presenting to the reader the account of Professor Peek, quoted in the "Introduction to Entomology:"—"This instrument is a very curious object; in order to describe it, it will be proper to compare it with the tenon saw used by cabinet makers, which being made of a very thin plate of steel, is fitted with a back to prevent its bending. The back is a piece of iron, in which a narrow and deep groove is cut to receive the plate, which is fixed. The saw of the Tenthredo is also furnished with a back, but the groove is in the plate, and receives a prominent ridge of the back which is not fixed, but permits the saw to slide forward or backward as it is thrown out or retracted. The saw of artificers is single, but that of the Tenthredo is double, and consists of two distinct saws with their backs. The insect, in using them, first throws out one, and while it is returning pushes forward the other; and this alternate motion is continued till the incision is effected, when the two saws, receding from each other, conduct the egg between them into its place. In the artificial saw the teeth are alternately bent towards the sides, or out of the right line, in order that the fissure or *skerf* may be made sufficiently wide for the blade to move easily. To answer this purpose in some measure, in that of the Tenthredo the teeth are a little twisted, so as to stand obliquely with respect to the right line, and their point of course projects a little beyond the place of the blade without being laterally bent, and all those in each blade thus project a little outwards. But the skerf is more effectually made and a free range procured for the saws, by small teeth placed on the outer side of each, so that while their vertical effect is that of a saw, their lateral effect is that of a rasp. In the artificial saw the teeth all point outward (*towards the end*) and are simple, but in the saw of the Tenthredo they point inward or towards the handle, and their outer edge is beset with smaller teeth which point outward (*towards the end*)."

When the little Sawfly has completed her incision (which, according to the species, is made in various parts of plants, in the stem, in the ribs of the underside of the leaf, or in the edge of the leaf), the egg is passed down, as has just been said, between the saws into the place thus prepared for it. Now it is well known that all wounds caused by a rough or blunt tool are more difficult to heal than those which are "clean cut." This holds good in the vegetable as in the animal subject, and it is here probably that the final cause of the complicated structure of this beautiful little instrument may be sought. It is not desirable that the wound should heal. The fissure in which the egg is inserted is not a mere resting place, but is designed to afford *nourishment* to the eggs, which, absorbing the juices of the plant, actually *grow* between the time of their exclusion and their hatching.* A supply of nourishment is thus produced and maintained by the stoppage of circulation consequent on the opening of this wound, which, in some cases, is further irritated by the introduction, at the time of oviposition, of a drop of poisonous fluid. In some cases this results in the formation of an excrescence similar to that produced by the Gallflies, within which the egg lives, grows, and is

---

* This phenomenon is not confined to the eggs of the Tenthredinidæ, it has been observed in those of the Ant; nor even to those of insects, as it is the case with fishes.

hatched. The red and green swellings so common on the leaves of various willows, and woody excrescences found upon the stems, are the production of species of the Tenthredinidæ.

It has been said above (p. 154) that while the larvæ of the rest of this order are footless grubs, incapable of seeking their own food, or in any way providing for their own wants, the larvæ of the Tenthredo family are active, and are left to "make their own living." In form and general appearance they closely resemble the caterpillars of moths. Like them, they have the six true legs (*i.e.*, legs which answer to those in the perfect insect), and, in most cases, the "false legs." These false legs differ from those of the moths, in being without *the circle of hooks* with which the false legs of the Lepidoptera are furnished. The false legs of the Tenthredo caterpillar differ in number also from those of the Lepidoptera, which never have more than three pairs of true legs on the thorax, four pairs on the abdomen, and one pair at the tail. The Tenthredo has three pairs of true legs, five, six, or seven pairs of abdominal legs, and one pair at the tail, except in two genera, where the false legs are altogether wanting. The examination, therefore, of the legs will at once enable the student to distinguish between the Hymenopterous and the Lepidopterous caterpillar.

The larvæ are chiefly green, sometimes yellow, with spots and tubercles, and have many of the habits of true caterpillars, for instance, of coiling themselves up, feigning death when alarmed, throwing out a silken rope by which to descend from one branch to another, &c. Some roll up the leaves of trees, and fastening them in that position by means of silken threads, thus construct a home closely resembling that of many caterpillars.

Others again, like the Clothes-moth and other casebearing caterpillars, form a little spiral case in which they move and eat as safely as a snail in his shell. These cases strongly resemble some made by the larvæ of some Caddis-worms.

These solitary cells are not the only dwellings formed by Tenthredo larvæ; some of the social species construct large silken tents, and it is not an uncommon thing to see a whole bush, or even a large portion of a hedge, almost covered with these silken webs, every twig denuded of its leaves and bound to those around it by innumerable little cables.

Some species have a curious property (which, however, is not confined to these insects) of emitting a fluid either from the mouth or from openings in the sides, which in some cases has an extremely disagreeable odour.

The larva of one species, *Selandria cerasi*, common on our pear, plum, and cherry trees, appears to be always enveloped in a dark slimy secretion, which so covers the insect as entirely to conceal it; in the words of Mr. Westwood—"The insect has not the least appearance of animation, and looks more like a small portion of slime."

The change of the Tenthredo usually takes place in a silk-lined earthen cell or cocoon in the ground; some, however, burrow into the pith of the stem of plants, while others construct a hard cocoon upon the twigs of trees. A large oblong or long ovoid cocoon of this kind is made by *Trichiosoma lucorum*, and is not uncommon on the thorn. When enclosed in the cocoon, the larvæ of some Sawflies remain for a considerable time before changing into pupæ.

The Sawfly is not without a place amongst the insects remarkable for maternal affection. It is stated that a Sawfly found in Van Diemen's Land sits on the leaf on which her eggs are placed until they are hatched, after which, like the Earwig, and the Plant-bug of the birch-tree, described by Dr. Geer, she guards them as a hen guards her young, covering them with her body and protecting them from all assailants. This exhibition of maternal feeling appears more remarkable in the Sawfly than in the other insects mentioned, on account of the far greater disparity of constitution, and almost, one might say, of nature, between the mother and the young. Both in the Plant-bug and the Earwig there is a great resemblance in habit, mode of feeding, and external form between the larva and the perfect insect, whereas in the Sawfly it is difficult to imagine any sympathy existing between the winged fly and the sluggish, crawling grub. We should smile at the idea of a white butterfly covering and tending a family of fat, green caterpillars; yet here is a phenomenon presented to us of a precisely similar nature.

The British species of the Tenthredo family number about three hundred.

Division II.—THE BORERS.—The Wood-borers (called *Sirex* by Linnæus, which name, like *Tenthredo*, is now applied to only a single genus of the family), very nearly resemble the Sawflies in general form and appearance; a variation in the form of the thorax, and the solitary spur on the fore-leg being the most conspicuous external difference. The tongue, as has been said (p. 156), differs from that of the Sawflies in being simple or undivided, while in the ovipositor we find the chief character which divides the present from the former family. The antennæ are thread-like, or very slender and bristle-like. They possess from ten to twenty-five joints, and the number of joints sometimes varies in the sexes.

The difference between the ovipositor of the Sawfly and that of the Woodborer presents an instance of the modification of structure to serve an especial end. It is impossible to examine this instrument in the two insects without perceiving that correspondence between them which would be looked for in animals so nearly connected, but the appearance of the two instruments is widely different, that of the Sawfly consisting chiefly of two thin and gracefully curved serrated blades, while that of the Borer is apparently a single strong boring implement. This auger, however, is found upon microscopic examination to be a kind of sheath, embracing, though not entirely enclosing two stiff serrated bristles which play, as it were, within the borer, and can be partially protruded. These latter are in fact the parts corresponding with the blades of the saws in the Sawfly, while their cylindrical case represents the backs of the saws in that insect. These backs, which in one case strengthen and support two independent saws, in the other, *soldered together*, form the principal part of the boring instrument.

The purpose of the variation in the ovipositor requires no explanation when the habit of the Woodborer is known. Not as with the Sawfly, in tender leaves, fruit, or the soft stems of plants, are the eggs of the Sirex to be deposited, but in the substance of sound and solid wood. Indeed, an account is given in the "Zoologist" (5829), taken from some French papers, in which numbers of *bullets* are said to have been found with circular holes drilled by the *Sirex juvencus*. Whether, however, these were bored by the ovipositor of the female, or were eaten out by the small strong jaws of the larva, is not stated; nor is any suggestion offered as to the purpose to be served in either case.

The Fir is especially subject to the attacks of the Woodborer, but it is to be found upon several other trees, as the Willow and the Hornbeam.

One large species, the *Sirex Gigas* (see Pl. VI. fig. 2) is well known in England, rather owing to its conspicuous size than to the frequency of its occurrence. It can hardly be called a British species, being probably imported in the larva or pupa state in foreign timber, and, so far as is known, not multiplying here.

The larva is an eyeless grub, the legs are only six in number, the insect being destitute of the "false" legs found in Tenthredo, and the body ends in a horny prickle.

The final transformation is said to take place within the wood excavated by the larva, after the formation, by the larva, of a "silken cocoon mixed with chips and excrement."

Only about ten species of Woodborers are known in England.

CHAPTER XIV.

HYMENOPTERA.—TEREBRANTIA.

Subsection II.—*ENTOMOPHAGA*.—We now come to the second subsection of the Terebrant Hymenoptera—Entomophaga. This consists of the Gallflies, Ichneumons, and Ruby-tails. All the insects of this subsection (with the exception of some of the Gallfly family) are parasitic. And here let it be observed that the Hymenopterous "Parasites" are by no means parasitic in the same sense in all cases. Some deposit their eggs in the bodies of other insects, most commonly when these are in the larva state. Here the eggs are hatched, and here the young prey upon the living substance of the unfortunate victim which feeds but to nourish "the wolf" inside, living only till its unwelcome guests are ready for their change; the caterpillar then either shrivels and dies or changes into a chrysalis, whence issues, not a moth or butterfly, but the host of little creatures which have been nourished on its embryo.

Another kind of Parasite merely deposits its eggs in the nests of other insects, where the larvæ feed on the provisions stored up for the young of the rightful owners of the nest. There are several genera of bees which are parasitic in this sense.

The present subsection, Entomophaga, presents instances of both these kinds of parasitism: it contains two divisions—

I.—*SPICULIFERA*, or dart-bearers, consisting of the Cynips family or Gallflies, and of the Evania, Ichneumon, Chalcis, Proctotrupes families; *parasitic on living insects.*

II.—*TUBULIFERA* (tube-bearers), consisting of the Ruby-tails, which are supposed to be *parasitic in the nests* of other Hymenopterous insects.

The *SPICULIFERA*, which are sometimes called *Piercers*, in contradistinction to the sawing and boring insects of the former subsection, like them derive their name from the nature of their ovipositor, a needle-like organ consisting of a horny sheath, guarded by a pair of valves and enclosing two slender and delicate serrated bristles.

The insects of this division differ widely in appearance from those of the two former, and the difference is that which we so frequently observe between carnivorous and herbivorous quadrupeds. The greyhound and the sheep—the tiger and the cow, do not present a greater contrast than we find between the substantial, straight-sided, oblong bodied Sawfly, and the light, almost fantastically formed little parasite Evania, or Chalcis.

It may be objected that this comparison is fanciful, and points to a relation which does not exist, because the larvæ, between which the diversity of food is found, are in both cases comparatively inactive grubs, while the perfect insects, which present the diversity of form, are almost all vegetarians. The answer to this is, that although one cause of the heavy form of the ox as compared with the tiger is that a larger bulk of vegetable than of animal food is necessary for nutrition, and therefore more room is required for its reception, this is not all. Carnivorous animals are mostly *predaceous*, and thus require an agility in motion and lightness of form unnecessary to the vegetarian. Now in the case before us this applies only to the perfect insects, which, notwithstanding that their victims are usually only larvæ, and sometimes stationary and even enclosed larvæ (as the larvæ in galls), are yet often put to a shift either to "catch their hare," or, having caught him, to seize the lucky moment for the achievement of the one work of their life. In fact, the very objection only serves to point out the economy more remarkably; this, which we may call the predaceous character of form, being given not to the individual which is to enjoy the advantage, but to that which is to secure it for him, the plan being thus carried out into two generations. That this is not an entirely fanciful idea may be presumed from the same circumstance occurring in the next section also. A very slight glance at the genera of bees (in the second section) will show that the lightest forms are to be found amongst those which are parasitic; whilst in the present division some of the heaviest forms in the Cynips family are to be found amongst the gall-makers—insects with vegetarian larvæ.

In these insects the head is small or of moderate size. The thorax is usually large in proportion to the abdomen, which, however, is often of great length. The abdomen varies greatly; in all it is attached to the thorax by a small point, which is sometimes drawn out into a long stalk, though in a few species (of Chalcis) some care is, in observing, required to avoid the mistake of supposing the abdomen to be attached as in the Sawflies. In one species it is most disproportionally small, whilst in others it is very long, compressed, and largest at the end; in some it is cylindrical, in others ovate or conical.

The legs are long in most of the parasitic families, less so in the Gallflies; many are of remarkable forms.

The wings vary—some species being altogether without wing-nerves, while in others these are well developed.

The Spiculifera approach more nearly than the Sawflies and Borers, to the predatory tribes contained in the next section, Aculeata; but, as has already been said, the legs afford a certain test as to which of these sections a species belongs to, the trochanter (p. 35) in all the Terebrantia consisting of two joints, while in the Aculeata it consists but of one. The antennæ, the wings, and the size of the insects also afford means of distinguishing them.

Thus in the Aculeata the antennæ are almost constantly twelve-jointed in the female, thirteen in the male; both fore and hind-wings are always veined. The insects are mostly of moderate or large size, the smallest seldom being less than $\frac{1}{4}$ or $\frac{1}{8}$ of an inch.

In the Spiculifera the number of joints in the antennæ is as follows. In the females of the Cynips and Evania families from thirteen to fourteen or fifteen; in the Ichneumon family it is generally above sixteen. These therefore may be distinguished by the antennæ. In the Chalcis family and the Proctotrupidæ the antennæ vary from six to sixteen joints, but in these the wings afford a sufficient distinction, the forewings being nearly or quite veinless, and the hind-wings entirely without veins. The extremely minute size of these little creatures is also in most instances sufficient to separate them from the Aculeata, only the giants attaining to the length of a quarter of an inch.

The Gallflies are the first family in this division. Known perhaps in their own little persons to naturalists alone, their works are familiar to all. The large oak-apple, so carefully sought for by children, to be covered with leaf-gold in loyal preparation for "King Charles's Day," is universally known. The little clear, globular red and green gall, hanging on its long strings from the catkins of the oak, and tempting to a parody of the old North country ballad—

> "O far hae I ridden,
> And meikle hae I seen,
> But *currants upon oak trees*
> Afore I ne'er saw nane ;"*

the "Artichoke gall," on the same tree; the round smooth gall so common now, though unknown in England forty years ago; all these, and many more, are among the common sights seen in every country walk by the least observant.

Other gall, there are, some as conspicuous as these, some even more so, which are less universally recognised as animal productions. The beautiful mossy tuft of crimson and green found on the stems of the wild rose, the small flat scales which sometimes entirely cover the under side of oak-leaves, some of the woody excrescences upon the trunks of trees, are alike produced by the gall-fly. But the young entomologist must not hastily conclude that all similar excrescences or morbid vegetable growths are the work of the Gallfly; many other insects, as the Sawfly (above mentioned), the Aphis, certain two-winged flies, and some beetles produce them; while certain morbid growths are produced by other agents. Thus,

---

* A large oak-gall growing near the Dead Sea has been seriously believed to be a species of fruit "which turns to ashes in the mouth ;" still bearing testimony till the present time of the sentence pronounced upon the accursed city.

for instance, the tops of nearly all the twigs in a quickset hedge may sometimes be observed to be thickened, rough, and apparently pierced with small holes, presenting an appearance very like that of old and deserted woody galls, but arising in reality from a fungous growth. So also an accidental injury to part of a tree will often produce a gall-like excrescence either on the wood or in the leaf-buds.

These galls are amongst the most puzzling of natural phenomena. All that is actually known is that the parent insect punctures stem, leaf, bud, or stalk, and there deposits an egg and (it is supposed) a drop of irritating fluid. It is not difficult to imagine that this, by arresting the circulation, might result in the formation of a shapeless or perhaps globular tumour; and in plants having a tendency to produce hairs, prickles, &c., that the tumour might be hairy or prickly; nor even might we see much difficulty in a modification of form or character depending upon the part of the tree affected. But what are the facts? On the leaf of the oak we find a small globular, smooth, clear gall, closely resembling that which is found growing on the flower-stalk of the same tree, and that which is found on the rose-leaf; and on the very same leaf we find a number of flat or slightly conical scale-like galls, covered with tufts of hair, and attached to the leaf by a short footstalk. It is clear that the oak-leaf has not insisted upon one mode of developing its little tumour. The rose-tree, upon the leaf or leaf-stalk of which we find the little smooth gall like that of the oak, examined farther, presents us with a ball of moss produced by the puncture of another Gallfly, proving that no necessity exists in the rose forcing it to develope this mossy covering wherewith to clothe all the excrescences which may be formed upon it; while an egg laid in one leaf-bud of an oak-tree results in a formation resembling an artichoke, and another egg laid in another bud produces a perfectly smooth, hard, round ball. Is then the cause of the difference in the fly and not in the tree? So far as has yet been observed each species of insect has its own form of gall, but this in no degree lessens the difficulty; for even supposing a chemical difference in the poisonous secretions of the various species, it is altogether inconceivable how so minute a drop as that to be deposited by an insect under the $\frac{1}{10}$ of an inch in length, and in so minute a wound, should occasion any serious disturbance at all of the circulation; but it is still more so that some difference in its composition should so regulate the whole process of change in the natural action of the tree, as to produce growths totally differing in appearance and in character. Conjectures have been formed to explain the whole of this process, but, like conjectures upon some other subjects, they are quite as puzzling as the original problem..

It is not only in abstruse matters of physiology that these little productions manage to baffle the naturalist. It is often no easy matter to discover or determine the owner of a gall from which *a tenant* has emerged. An amusing instance of this occurred in the case of Réaumur. Wishing to witness the growth of the mossy rose-gall, he carefully tended a number of flies lately hatched from one, supplying them with a branch from a rose-tree, in order that they might lay their eggs therein and prepare the way for future galls. After waiting for some time, and finding that the flies showed no disposition to attack the rose-branch, he discovered his little brood to be not Gallflies, but *Ichneumons* (insects also belonging to this division, and to be described hereafter), which had been deposited as eggs *within the Gallfly young*, and, having lived upon their substance, were matured and came forth in their stead. Another instance occurred to the present writer, who, in 1857, found upon the leaves of a wild rose-tree at West Wickham a new and very beautiful little gall, nearly globular, and crowned with spines. This was sent to the British Museum, and from it were hatched two species of Spiculiferous insects. These were supposed to be the Gallfly and its parasite; and the question was, "which was which?" The answer was "neither," for the insects *both* proved to be parasites—one probably on the Gallfly larva, the other on the Gallfly's parasite; so true it is that "big fleas have little fleas upon their backs to bite 'em."*

Other tenants may be met with in galls; Mr. Stainton (in the "Zool." 5139) mentions finding the caterpillars of moths in the mossy rose-gall;† and also its being a known fact that another moth larva is bred from oak-apples in Germany; while Mr. Walker enumerates about twenty-five species of insects of the orders Coleoptera, Orthoptera, Neuroptera, Diptera, Lepidoptera, and *Hemiptera*, besides five or six species of spiders or acari, which emerged from oak-galls under his observation in one year. These, however, are probably instances of mere cuckoo-like parasitism, as it is not likely that these caterpillars feed on the young gall insects.

---

* It may be worth recording, that this gall having been sought in vain for ten years in other places, was looked for on the same rose-tree in 1867, and again found there.

† From this gall, at least six species of Spiculifera (including the owner) have been reared.

The largest Gallfly known in England is that which produces the round hard gall now so common upon low, or young oak-trees, and spoken of above as unknown thirty or forty years ago. This insect (Cynips quercus Lignicola, Pl. VI., fig. 3) was at one time believed to be identical with that of the Aleppo Gall, the gall of commerce, and to have been imported in the gall, afterwards naturalizing itself in England. It is now, however, recognised as a *distinct species*, and the origin of its introduction is as obscure as ever. For some time attempts were made and renewed to utilize this Gall in the manufacture of ink, but the tannin, or dyeing matter, yielded by it is so inferior in quality to that of the foreign species, that it seems unlikely that it should ever supersede this. These Gallflies were, with those of another species, made the subject of a curious experiment by Mr. F. Smith ("Zool." 7330). The male fly of the Cynips being unknown, he collected about a bushel and a half of the Galls, with the purpose of discovering it, if possible. From all this number none but females emerged. These he placed on oak-trees in various places, and afterwards visiting them found new Galls upon those trees, and on no others in the neighbourhood.

From another species of Galls he obtained about 1200 Flies, all female, whilst Hartig, as Mr. Smith mentions in his paper, amongst 10,000 of one species and 4000 of another, could not discover a single male.

The Gallflies are often rather heavier in figure than most other of the Spiculiferous insects (see Pl. VI. fig. 3), with which, however, they are very nearly allied, not only in structure but even in habit, some species having been discovered to share in the parasitic habits of those other families—laying their eggs in the bodies of other insects.

The head in the Cynipidæ is small, the thorax thick and oval. The abdomen, largest towards base and egg-shaped, is sometimes (but not always) much compressed, and is attached to the thorax by a stalk, which is in most cases very short. The antennæ are slender in the male, less so in the female (let it be observed that the absence of the male is remarked in the genus Cynips, not in the whole family of the Cynipidæ), and the legs are of moderate length. The wings vary, and are very long in some species; they have but few veins. The female deposits her eggs either singly, or many together, of which the two large oak-galls are examples; the larger, found on full-grown trees, containing many inhabitants, while the round smooth hard Gall, lately described, contains but one.

We now come to the true Spiculiferous parasites.

The first family, Evaniidæ, contains one of the most whimsically proportioned insects in the whole order, Evania appendigaster, supposed to be parasitic upon the small Cockroach. This little insect has an enormous thorax, a smallish head, long legs (especially the hinder pair), and an abdomen so small that it seems impossible that it should belong to the insect.

This family contains but three British genera, and only about half a dozen species. It may easily be distinguished from the following by the attachment of the abdomen to the thorax, not at the apex of the latter, but *from its upper side*, giving it the appearance of springing out of the insect's back.

The next family, Ichneumonidæ, is far more numerous, containing about 120 genera, and more than 1100 species. It includes all the Ichneumon flies of large size, with the exception of one or two which belong to the Evaniidæ, and also some very minute species.

The Ichneumons are elegantly formed insects, combining the appearance of lightness and of strength, and with some of them, at least, the reader must be familiar. The wings are large and firm, and beautifully veined, forming, in the front pair, several perfect cells ; the head, which is of moderate size, is set lightly on a compact thorax, larger before than behind ; the abdomen, long and slender, sometimes much compressed and abruptly truncated, is set on by a small point, or sometimes by a fine stalk *at the extreme end of the thorax, between the hind legs.* The legs are of moderate length ; the antennæ are long, slender, and tapering ; the ovipositor is in some species short and concealed within the abdomen, in others it is visible and occasionally of great length, considerably exceeding that of the body. Here is a structural variety which at once points to a variety of habit, and accordingly we find that while some species deposit their eggs within, or upon the bodies of exposed and naked larvæ, others, by the exercise of the powers of smell, touch, or we know not what, discover the hiding-place of the larvæ most carefully concealed from sight and guarded from danger, and with their long ovipositors succeed in lodging their eggs within the bodies of the victims. Thus do some species penetrate to the little grub within the heart of the oak-gall ; others find the wild Bee in its cell, the Beetle in its wooden chamber hollowed out within the trunk of the forest tree.

Others, again, display a still more remarkable instinct, the perfect insect actually entering the water in order to deposit her eggs within the bodies of aquatic larvæ.

One of the best known of the Ichneumons is, perhaps, the Yellow Ophion (Pl. VI., fig. 4). Frequently attracted by the light through the open window in our rooms on a summer evening, it seldom fails to attract attention. Its active movements, its size (it is nearly an inch in length), the beauty of its large, clear, bright wings, and the noise which it makes in striking repeatedly against the ceiling, all render it difficult to overlook the little visitor. Another genus of flies resembling the Ophion is Paniscus, which may be distinguished from it by the presence of a very small triangular cell (the "second sub-marginal"), which is wanting in the wing of the Ophion. The Ophion is also to be distinguished by its beautiful comb-like foot-claws.

The Ophion differs from most of the Ichneumonidæ in depositing her eggs, not within the body of her victim, but upon its surface. According to Kirby and Spence, the egg is curiously attached to the body of the Caterpillar by a short footstalk, which is fastened into the skin by an enlargement of the lower end, like the root of a hair. The Ophion larva, when hatched, does not quit the egg, but, keeping its hinder end within the broken shell, and laying hold of the Caterpillar with its jaws, remains feeding in this position till the time for its change. So firmly fixed is it that not even the moulting of the Caterpillar dislodges it.

The greater part, however, of this family lay their eggs within their victims, which are usually the larvæ of insects, sometimes the perfect insects and sometimes the eggs. The benefit thus conferred by the Ichneumons upon man is inestimable. They appear to be the principal means employed to check the devastations of the vegetable-eating tribes, which would otherwise lay the country bare of food, and it has been remarked that in those years in which any one species of Caterpillar has been unusually abundant, the Ichneumons have been proportionately so. Some years ago, the speckled Caterpillars of the Currant Moth were so abundant at Bognor, in Sussex, that it was almost impossible to walk without crushing them by hundreds. The roads were full of them; the houses were full of them; trees, palings, walls, were covered by them: it was rare to see a few square inches without one or more of these little animals. A woful prospect for the following year, if all of these—if, indeed, an average proportion of these—should come to maturity, and each one should lay its hundred or so of eggs, to be developed into as many more hungry Caterpillars! But what happened? In a few days, trees, walls, palings, were covered with clusters of beautiful little yellow silken cocoons, each containing the germ of a little Ichneumon—one of a numerous family which had been feasting within one of these larvæ; and that year the Currant Moth was hardly more abundant than usual. And what became of the little Ichneumons? Possibly, in their turn, they fell a prey to others—as in this family it is not unfrequently that parasite preys on parasite; perhaps some other animal was made happy by an unusual supply of food. Anyway, we may be sure that these myriads of little creatures were not called into being without a proportionate amount of enjoyment in the world; that their lives were not wasted; that their death was but a means of supplying with life and enjoyment yet another race of living beings.

Some of the Ichneumons deposit but one egg in one

victim, others deposit more than a hundred. Some undergo their change within its pupa case, while others desert the dying caterpillar, which in this case undergoes no metamorphosis.

An idea may be formed of the minute size of some of the species, if the student will examine the first twig of a rose-tree that he can find. On this he will see what at first appear to be dead Aphides, brown, hard and stiff, but retaining their perfect form, and, in reality, consisting but of empty skins. In each of these he will observe a small round hole, and out of this hole he may know that a little Ichneumon Fly has emerged—having been born and brought up within the body of the Aphis. It is said that a Chalcis is parasitic upon this parasite. Bonnet (quoted in "Insect Transformations") tells of a "prodigious number" of some of these parasites being hatched from 20 *butterflies' eggs*.

It is a curious fact that many insects appear conscious that the Ichneumon Fly is their natural enemy, show fear at her approach, and endeavour to elude her attacks.

The fourth family consists of the Chalcis and its relations (see Pl. VI., fig. 5, C. flavipes). These are mostly very minute parasites of beautiful metallic lustre and colouring. Their wings have but few veins; in the minutest species none at all. They do not differ greatly from the former family in their habits, but exhibit some peculiarities of structure. The most conspicuous of these is that in many species the hind legs are of an extraordinary form, the femur or thigh being enormously thickened and sometimes toothed, while the unusual appearance is increased in some species by a great length in the coxæ of the hind legs.

Here, again, is a structure which leads to inquiry, and our experience of the large-thighed beetles naturally leads us to expect great power in leaping in these little flies. But what is the fact? That many of them possess no such power, and that we find no reason whatever for this extraordinary development. The lesson, therefore, which we have learned from the little Chalcis, is not that we may safely presume upon our experience to jump at conclusions, but that when we feel most certain beforehand of how our natural history facts *ought* to turn out, we had need to be most careful to ascertain whether they may not prove exactly contrary.

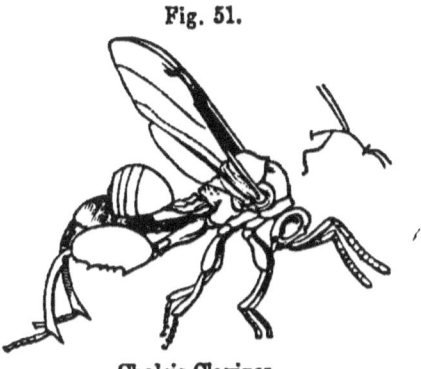

Fig. 51.

Chalcis Clavipes.

Some of these little creatures are parasitic upon other parasites, whose presence they discover whether on the exposed body of a naked insect or in the grub enclosed within a gall; many prey upon the gall grub itself. One of these, *Callimome flavipes*, found in the round hard oak-gall, is described by its discoverer, Mr. Parfitt, in terms which present an image to the mind only less gorgeous than Blake's vision of the green-mailed ghost of a flea, holding its golden cup of blood. "Wings splendidly iridescent; head, thorax, and abdomen beneath of the most magnificent shining green; the basal and two next segments of the abdomen very highly polished, and reflecting a steel-blue in certain lights; eyes brown, &c. The length of this glorious atom is about a quarter of an inch."

The Chalcididæ number upwards of 1190 species, and are of great value in keeping under the numbers of injurious insects. The "Death-watch," which feeds on the woodwork of houses, is among their victims.

The fifth and last family of the Spiculifera is Proctotrupidæ, or the Proctotrupes family. These, like Chalcididæ, are very small insects, with few or no veins in their wings. They differ in form from Chalcis, which is the most *squat* of the parasitic families, being of more slender proportions. They may also be distinguished from these by their antennæ, which are less decidedly angulated, and by the absence of a hollow in the forehead, which in Chalcis receives the antennæ. In colour they are less conspicuous, being chiefly black and brown, and although in some species the thighs are slightly thickened, they never present the remarkable appearance of the legs in some of the Chalcididæ.

If the small size of the Chalcididæ was shown by their being hatched within an aphis, or a butterfly's egg, it may give some idea of the minuteness of the Proctotrupidæ to mention that the little caterpillar which mines the roseleaf is said by Mr. Westwood (from Dr. Gees) to be infested by one species. Many individuals have also been reared from one butterfly's egg, and they have been found in the larvæ of a minute insect, which feeds within the envelope of a grain of wheat.

There are about four hundred British species in this family.

Division II.—*TUBULIFERA.*—*Tube Bearers* or *Ruby Tails* (see Pl. VI. 6).—The Chrysis, or "ruby tail," is a beautiful insect which can hardly have failed to attract the attention of the reader. The fretted surface of the head and thorax is a deep but brilliant green or blue, the abdomen is crimson or a glowing coppery red, with sometimes a golden lustre: The wings and legs are rather small, the proportion of the whole insect being somewhat similar to that of many of the small wild bees, while the resemblance is increased by the occasional aggressive use of the ovipositor, which, however, having no poison-bag attached, cannot be considered as a true sting. The antennæ also resemble those of the bees in being decidedly *kneed*. The Chrysis has a curious habit, by which alone it might be recognised, of rolling itself up when annoyed or alarmed into a little sting-proof ball, of almost as perfect a shape as that assumed by the many-jointed woodlouse, though in the Chrysis this form is attained by merely turning down the abdomen (which is very convex above and concave below), until it fits closely upon the headside of the thorax, leaving the wings sticking out, unprotected, in a straight line. In this position it looks like a little jewel, half ruby, half emerald.

The chief structural peculiarity of the Chrysis, and that from which it derives its name, is in the abdomen. This appears to be formed of no more than from three to five (according to the genus and the sex) large segments, the remaining segments being apparently but part of the ovipositor. They form small tubes, which can be drawn into one another like the joints of a telescope, and are terminated by an ovipositor resembling in its main characters those of the preceding division.

In habit the Chrysis is parasitic, not as the Ichneumons, &c., but depositing its eggs in the cells formed by the wild bees, sandwasps, and even into the long deep tunnel of some solitary species of true wasp. Here the larva feeds either on the stores there laid up, or, as seems probable, on the young owners themselves of the cells.

The female may frequently be seen running busily about on posts and palings such as the Carpenter Bee delights to choose for her nest, or on sandy banks; in short, wherever her victims may be found. Carefully watching her opportunity, she seizes the right moment for depositing her egg or eggs, notwithstanding a spirited resistance which sometimes takes place on the part of the rival mother. Mr. Westwood relates—from M. le Comte de Saint Fargeau—an amusing instance of this. One of the Mason Wasps, "returning to its nearly-finished cell, laden with pollen-paste, found the Hedychrum (one of the Chrysis family) in its nest, which it attacked with its jaws. The parasite, however, immediately coiled itself into a ball, so that the Bee was unable to hurt it. The Bee, however, bit off the four wings, which were exposed, rolled it to the ground, and then deposited its own load in the cell and flew away. Whereupon the Hedychrum, now wingless, had the persevering instinct to crawl up the wall to the nest, and there quietly deposit its egg, which it placed between the pollen-paste and the wall of the cell, which prevented the Bee from seeing it."

There are five genera and twenty-four British species in this family.

CHAPTER XV.

HYMENOPTERA.—ACULEATA.

THE second section of Hymenoptera, ACULEATA, derives its name from the character of the ovipositor. This, aculeate, or needle-like, as it appears, is but one more modification of the same parts as are found throughout the other section of the Hymenoptera, consisting principally of two fine serrated bristles, enclosed in the horny duct which gives the name of aculeus to the instrument. In this tribe, however, the ovipositor is connected with a bag of poison, and to this peculiarity it owes the especial name of *sting*.

To HYMENOPTERA ACULEATA, then, belong all the true stinging insects, Ants, Wasps, and Bees.

The Section ACULEATA is divided into two Subsections, named from the habits of the insects—(1) *PRÆDONES*, or the Rapacious, (2) *ANTHOPHILA*, or the Flower-loving Hymenoptera.

The *PRÆDONES* consists of the Ants, Sandwasps, and true Wasps.

The *ANTHOPHILA* consists of the Bees.

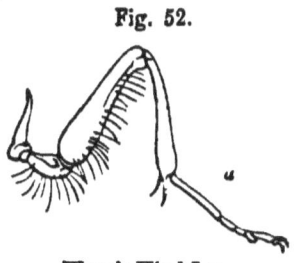

Fig. 52.

Wasp's Hind Leg.

The insects belonging to these two Subsections may be distinguished by the legs. In the Prædones, the first or basal joint of the hind tarsus (fig. 52, *a*) is cylindrical, while in the Anthophila, it is enlarged, and more or less *flattened*.

188   INSECTS.

This peculiarity in the Bees will be further noticed in its own place.

Subsection 1.—*PRÆDONES.*—The Prædones (*Prædo, a robber*) form three Divisions — 1. *Heterogyna*, containing the Ants; 2. *Fossores* (*Fossor, a digger*), containing the Sandwasps; 3. *Diploptera*, containing the true Wasps. The two first are easily distinguished from the third, *which has the fore-wings folded length-*

Fig. 53.

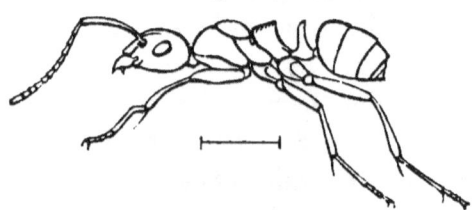

Outline of Formica.

ways when at rest (whence the name Διπλόω, *diploo*, to *double;* πτερὸν, *a wing*). The other two divisions are less easy to distinguish from each other, but the following rules may be sufficient. Heterogyna consists of the Social Ants and of the Solitary Ants. The Social Ants may be distinguished from the Sandwasps by the form of the footstalk by which the abdomen is attached to the thorax, and which forms, in some, one scale-like projection (fig. 53), or a knot-like lump; in others (fig. 54), two such lumps or nodes.

The Solitary Ants may be distinguished from the Sandwasps by the female of the former being always wingless, while the male is generally toothed or spired at the apex of the abdomen.

When, therefore, an insect has been shown by the legs to belong to the subsection *PRÆDONES*, and by its nonfolding wings to belong to one of the two first divisions in that subsection, its place will be further ascertained

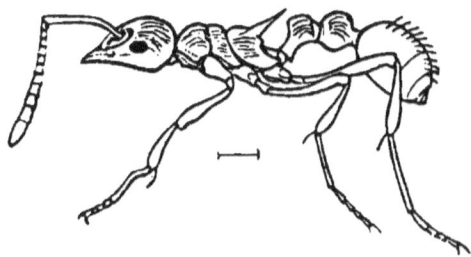

Fig. 54.

Profile of Myrmica.

thus. It belongs to the first division, *Heterogyna*, if, whether wingless or winged, the abdomen be furnished with the scales or nodes (social ants); also if it be wingless (♀ of social ants sometimes, of solitary always); also if it be winged and spicate at the tip of the abdomen (♂ of solitary ants only). Otherwise the insect belongs to the Sandwasps or Fossores.

Division I. — *Heterogyna.* — *The Ants.* —Very different opinions are entertained as to this division *Heterogyna*. The word, signifying ἕτερος, *heteros*, different; Γυνὴ, *gyne*, a woman, is by some considered to exclude the *mutillidæ* or solitary ants, in which the sexes consist only of the perfect male and female. The female here, however, differs from most perfect insects in being *always wingless*, in this approaching the neuters of the social ants. In adopting the present arrangement, the writer follows Mr. F. Smith, of the British Museum, who has retained it from older writers, and this is done partly in the hope that the young student of Hymenoptera may, as a first step, possess himself of the delightful (so-called) "catalogues"* by this gentleman, published by the British Museum, and proceed upon his studies with these in hand.

Heterogyna, then, consists of the Social Ants, *Formicidæ*, now divided into *Formicidæ*, *Poneridæ*, and *Myrmecidæ*, and of the Solitary Ants, *Mutillidæ*.

The social ants are distinguished from the solitary and from all other hymenopterous insects by a peculiar development of the first, or first and second joints of the abdomen. The first joint, which forms the *stalk* of the abdomen, grows out behind into a scale or raised "node," in the Formicidæ and the Poneridæ (fig. 53, p. 188); in the Myrmecidæ (fig. 54, p. 189), the same happens with both the first and second joints.

The social ants, like the social wasps and bees, consist of males, females, and workers, or imperfect females, the latter being *always wingless* among the ants, while among the bees and wasps they, like the perfect insects, possess permanent wings. The female of the social ants, winged like other insects at her emergence from the pupa state, and like them, rejoicing for a time in the sunshine and fresh air, to exercise them, retains her wings only until she is ready to enter upon the business of her life, laying the eggs which are to fill the nests preparing for them by the workers. She then prepares herself for her underground labours by voluntarily depriving herself of these appendages.

It would require more than the bulk of this entire volume to repeat the wonders recorded of the tribe of social ants. The very bees yield to them in the variety

---

* *British Bees*, and *British Fossorial Hymenoptera*.

of their interests and achievements. Not only are they the most skilful architects amongst insects, but as statesmen, as soldiers, as landed proprietors, as slave-owners, herdsmen—nay, if some writers may be believed, as *agriculturists**(!), they stand at the head of insect-thinkers and doers. Yet, above all, do they claim our sympathy and respect in one point as yet unnamed, that is, in their marvellous domestic conduct; their unceasing industry and tenderness in behalf of their young,— tenderness, not maternal, for it is shown by those who are not and cannot be mothers; while their private character is still further displayed by the friendship, good understanding, and care for the safety one of another, which subsists amongst the individuals of the community. An amusing case of somewhat officious and peremptory exercise of the privileges of friendship was observed by Hagen:—"The legs of a glass case, which contained the nest of social ants, were plunged into pans of water, to prevent the escape of the ants; this proved a source of great enjoyment to these little beings, for they are a thirsty race, and lap like dogs. One day when he observed many of them tippling away merrily, he was so cruel as to disturb them, which sent most of the ants in a fright to the nest; but some, more thirsty than the rest, continued their potations. Upon this, one of those that had

---

* It is said that a species of Ant in Texas actually plants around its dwelling a kind of grass, which it "nurses and cultivates with constant care, cutting away all other grasses and weeds that may spring up." Another species is said to plant "shade trees" as a protection against the summer sun. (See " Zool.," 7576.)

It is possible to believe almost anything of the Ants, but even "seeing" ought not always to be " believing."

retreated returns to inform his thoughtless companions of their danger: one he pushes with his jaws, another he strikes first upon the belly and then upon the breast, and so obliges three of them to leave off their carousing and march homeward. But the fourth, more resolute to drink it out, is not to be discomfited, and pays not the least regard to the kind blows with which his compeer, solicitous for his safety, repeatedly belabours him. At length, determined to have his way, he seizes him by one of his hind legs, and gives him a violent pull. Upon this, leaving his liquor, the loiterer turns round, and opening his threatening jaws, with every appearance of anger, goes very coolly to drinking again. But his monitor, without further ceremony, rushing before him, seizes him by his jaws, and at last drags him off in triumph to the nest."

After a battle, or any accident which has befallen a colony, the survivors invariably carry away their dead; in the case of a battle the conquerors carry away the bodies of their own soldiers, leaving the others to their fate.

The community in which these insects live consists primarily of the females and the workers, or "neuters," or imperfect females. The males and females leave the nest on arriving at perfection, and associate together for a short time; after which the males die, the females alone returning to the nest, where they labour diligently until sufficient workers are hatched to set them free from "menial" offices.

Their principal business from this time is the laying of eggs, which are received and tended with the greatest care by the workers. These carry them from the place where they are dropped and carefully deposit them in suitable chambers, moistening them, it is said, from their own mouths, and thus probably affording that nourishment which must be essential to their *growth;* the eggs of Ants, like those of Sawflies, growing larger after they are laid. According to the observations of M. Hubner, the nurses then bestow the most assiduous attention upon the eggs, daily removing them to those parts of the nest of which the temperature is most suitable. In the morning the eggs are carried to the upper chambers, to be within the influence of the sun's rays, while in the evening they are transferred to the lower apartments, which are less susceptible to a sudden lowering of the temperature. The eggs hatched, yet further labours devolve upon the careful and busy nurses, who to the daily removal of their little charges (creatures which before long are equal to themselves in size) now add the task of supplying them with food; or, rather, of feeding them. Nor does their care end here: when the time for its perfection arrives, the larva, having spun its own cocoon (the only act which it has ever been allowed to perform for itself), is not only extricated by the workers from its silken shroud, but even receives their assistance in divesting itself of the delicate membrane which still has to be stripped from its body.*

It has been said that the community consists *primarily* of the females and the workers, but this is not all. The

---

* That all this care is not absolutely necessary has been proved by the experiments of Mr. F. Smith, who found that the young ants, deprived of the assistance of their nurses, were able, in some cases, to emerge without help from their pupa-cases. Mr. Smith observes that the pupæ are not always enclosed in silken cocoons, the naked pupæ always giving out neuter insects. He accounts for this on the supposition that the under-fed female larvæ which were to be imperfectly developed into neuters, were not sufficiently nourished to produce the silk.

nests of Ants present the remarkable phenomenon of being inhabited by various other species of insects, concerning many of which there seems to be now no doubt that they are actually kept prisoners by the Ants to serve in various ways to the maintenance of the state. Amongst these are the Aphides, commonly called the cows of the Ants (whence we have given to the latter the name of cattle owners), species which feed on the roots of grass, &c., being plentiful in the nests, whilst others, Leaf-eaters, are sometimes enclosed by the Ants in a kind of earthen gallery constructed on the twig which forms their pasture. Numerous species of Beetles are also well known as inhabitants of ants' nests, and though it seems difficult to ascertain in all cases whether this is in the interest of the Beetles or of the Ants, yet in some there is no doubt that the Ants derive from the Beetles, as from the Aphides, a fluid which serves them as food. Mr. F. Holmes (*Zool.* 475) saw some large Red and Black Ants carrying as captives living specimens of Philonthus; while other observers have seen the ants forcibly preventing the escape of certain Beetles from the nest. Woodlice also are found in great numbers in the nests of ants, and whether or not these are amongst the profitable servants of the commonwealth, there can be little doubt that their residence would be but of short duration if disapproved by their omnivorous little hosts. With regard to the Aphides, and some of the Beetles, the question is put beyond a doubt by the sedulous care taken by the Ants of these herds—the eggs of the Aphides receiving attention equal to that paid to their own.

But yet another element exists in the community of some species of Ants—more or less warlike, as are all the social tribes, contending to the death for their territorial possessions; there are some whose taste in this direction is so *prononcé*, that they make war for the purpose of possessing themselves of slaves who shall free them from the necessity of all home drudgery. This, which might seem almost incredible, can scarcely be refused credence on the authority of such writers as those who have from their own observation described the proceedings of these slave-making Ants. The slaves once domiciled amongst their captors take willingly to their work, and perform most efficiently all the duties of builders, nurses, and housekeepers, even extending their labours to the feeding of those heroes whose inveterate laziness "off duty" is not without example amongst the warlike portion of a larger if not nobler race.

Mr. Newman, in his "Popular Introduction to the Natural History of Insects," gives a description of the proceedings of these Ants, which will serve to illustrate many points in their military tactics.

"The most remarkable fact connected with the history of Ants is the propensity possessed by certain species to kidnap the workers of other species and compel them to labour for the community, thus using them completely as slaves, and, as far as we yet know, the kidnappers are red or pale-coloured Ants, and the slaves, like the ill-treated natives of Africa, are of a jet black.

"The time for capturing slaves extends over a period of about ten weeks, and never commences until the males and females are about emerging from the pupa state, and thus the ruthless marauders never interfere with the continuation of the species. This instinct seems specially provided, for were the slave ants created for no other end than to fill the station of slavery to which they appear to be doomed, still even that service must fail were the attacks to be made on their nests before the winged myriads have departed, or are departing, charged with the duty of continuing their kind.

"When the Red Ants are about to sally forth on a marauding expedition, they send scouts to ascertain the exact position in which a colony of negroes may be found. These scouts having discovered the objects of their search, return to the nest and report their success. Shortly afterwards the army of Red Ants marches forth, headed by a vanguard, which is perpetually changing; the individuals which constitute it, when they have advanced a little beyond the main body, halting, falling into the rear, and being replaced by others. This vanguard consists of eight or ten ants only. When they have arrived near the negro colony they disperse, wandering through the herbage and hunting about as if aware of the propinquity of the object of their search, yet ignorant of its exact position. At last they discover the settlement, and the foremost of the invaders, rushing impetuously to the attack, are met, grappled with, and frequently killed by the negroes on guard. The alarm is quickly communicated to the interior of the nest; the negroes sally forth by thousands, and the Red Ants rushing to the rescue, a desperate conflict ensues, which, however, always terminates in the defeat of the negroes, who retire to the inmost recesses of their habitations. Now follows the scene of pillage. The Red Ants with their powerful mandibles tear open the sides of the negro ant-hill and rush into the heart of the citadel. In a few minutes each of the invaders emerges, carrying in its mouth the pupa of a working negro, which it has obtained in spite of the vigilance and valour of its natural guardians. The Red Ants return in perfect order to their nest, bearing with them their living burdens. On reaching the nest the pupæ appear to be treated precisely as their own, and the workers when they emerge perform the various duties of the community with the greatest energy and apparent goodwill; they prepare the nest, excavate passages, collect food, feed the larvæ, take the pupæ into the sunshine, and perform every office which the welfare of the community seems to require. In fact, they conduct themselves entirely as if fulfilling their original destination."—*Newman's Familiar Introd. to the Nat. Hist. of Insects*, p. 50.—(From the " Zool.")*

Slight as has been the preceding sketch of the habits and manners of the Ants, too many pages of this small book have already been bestowed upon them, and therefore but a few lines more may be devoted to the mention of their architectural labours. These are no less wonderful than their other proceedings, and the reader is referred to the pages of Messrs. Kirby and Spence for a most delightful *resumé* of, and observations upon, this and other of their achievements. Suffice it to say here, that without bricks and without mortar they build their many-chambered dwellings—build them of loose sand compacted apparently by some especial mode of manipulation. Story upon story of chambers are there connected by galleries and supported by pillars and buttresses, the nest being closed and guarded by doors, which are daily removed and nightly replaced. The edifices of various species vary in plan, and display the application of

---

* The above was transcribed some time ago, and the writer, not having the "Zoologist" at hand, is uncertain as to whether it was transcribed verbatim.

various architectural contrivances; such as the use in one case of *beams* in the construction of a ceiling, while in another a large chamber will be strongly roofed without beams or central support, by the application of the *arch*.

The nests here spoken of are constructed in the earth, those of some species are excavated in the trunks of old trees. Their internal temperature is high, the Ants, like the Bees, having the power of generating a considerable degree of heat. They are strongly redolent of a secretion peculiar to the Ants, formerly called "formic acid," and which is nearly powerful enough to take away the breath if the head be held over a large and disturbed nest. This acid the Ants have the power of squirting to a considerable distance, and it forms a considerable weapon in their warfare. The whole of the inside of nests hollowed out in the trunks of trees is stained black by this acid, while Ray records that blue flowers placed in an ant-hill turn red, and that a similar effect is produced in a Bluebottle by the sting of an Ant.

It is probably for the sake of this acid that the insect is used by the New Zealanders in the composition of the wourali poison.

In Switzerland Ants are used crushed into a plaster or poultice to be applied to the head to cure the headache, while their stimulating property is well known to Swiss schoolgirls, who rub their foreheads with the insects "*pour se fortifier la mémoire.*" Ants give out the acid so freely that in the same country the children lay a wet branch across the nest of the large Wood Ant, and when it is well covered with the insects, brush them off and suck from it the "hot vinegar."

The principal use, however, made of these insects, both in Switzerland and Germany, is in the composition of *ant-baths*, on the subject of which a friend of the writer has kindly communicated the following:—

"The first time I ever heard of the ant-baths was when at Wildbad. A Russian lady in the same house with us, after having taken a course of the baths there, was ordered to visit a village (of which I forget the name) in the heart of the Black Forest, to strengthen herself by the use of ant-baths. Afterwards, when at Wiesbaden, our landlady told us that these baths were very commonly used. Her own daughter, when a child, had derived great benefit from them. At five years old she could not walk, and had dwindled away to a mere skeleton, when the mother was advised to try ant-baths, which completely restored the child's strength.

"The ants are the large Wood Ants, and are collected, earth, stones, leaves, &c., all together in bags, which are placed in the bath, and have boiling water poured on them. This is left to stand some time, and the water is then used for the bath. They are sold in bags in the market at Wiesbaden at the proper season, and are used also for making ant spirit. For this purpose the ants are put into a glass bottle filled with some cheap spirit, and hung in the heat of the sun for some time. This spirit is used to rub the limbs in the case of sprains or weakness.

"Ant vinegar is made in large quantities every year by the Swiss ladies."

Possibly a liking for this acid is one of the attractions to some of those species of beetles which reside voluntarily in ants' nests, as they have been found inhabiting old nests, and deserting them when a heavy shower had washed away the acid.

Much discussion has arisen upon the often-quoted

words of Solomon: *Go to the ant, thou sluggard: consider her ways, and be wise: which having no guide, overseer, or ruler, provideth her meat in the summer, gathereth her food in the harvest.* It has long been popularly supposed that the ant does actually store up grain as food for winter use, and the resemblance to some small grain of the white pupæ so carefully laid up, so eagerly seized and carried away to some safe place on the disturbance of a nest, has fostered, if not given rise to this idea. The truth, however, is that, at least in England, the Ants spend the winter in a torpid state, neither requiring nor possessing magazines of food; the food so industriously collected at other times being for the immediate consumption of the inmates of the nest. That seeds of various kinds are collected by Ants and carried to the nest is beyond a doubt, but all observations point to the fact that these are used not as food, but as building material, in common with small stones and other small objects which are collected at the same time and in the same manner. Possibly in this fact may be found an explanation of the supposed agricultural performances mentioned in the note at p. 191.

How these facts are to be reconciled with the words of the inspired writer remains to be shown. Possibly a further knowledge of the habits of ants in warmer climates may do this, or possibly it may be a question for the Philologist rather than for the Hymenopterist, as it is by no means easy, nor always possible, to ascertain without doubt the exact species of animal to which the Hebrew names apply. In the volumes of Messrs. Kirby and Spence, however, the following remarks occur, and seem to remove the difficulty on a sound principle:—

"I think, if Solomon's words are properly considered, it will be found that this interpretation has been fastened upon them, rather than fairly deduced from them. He does not affirm that the ant, which he proposes to his sluggard as an example, laid up in her magazine stores of grain. . . . . The words may very well be interpreted simply to mean that the ant, with commendable prudence and foresight" (and surely we may add with industry), "makes use of the proper seasons to collect a supply of provisions sufficient for her purposes. There is not a word implying that she stores up grain or other provision. She prepares her bread, and gathers her food—namely, such food as is suited to her in summer and harvest—that is, when it is most plentiful, and thus shows her wisdom and prudence by using the advantages offered to her. The words thus interpreted, which they may bear without any violence, will apply to our European species as well as to those that are not indigenous."*

The Social Ants, formerly all included under Formicidæ, now form the three families, Formicidæ, Poneridæ, and Myrmecidæ. These are distinguished by the "nodes" on the abdomen. In the two former families there is but one, in the latter two (see figs. 53, 54, pp. 188, 189; and Pl. VII., fig. 1). The females of the Formicidæ present an exception in the Section to which they belong, being without a sting. In Poneridæ (of which there is but one, and that a rare English species), and the Myrmecidæ, both females and neuters are provided with this weapon.

The Solitary Ants, or Mutillidæ, although a very numerous family abroad, consist in England of but

---

* The word translated "provideth" does not necessarily imply *foresight*. In Gen. xliii. 16, the same verb is translated "make ready."

three genera, these containing only five species. They appear to be very nearly related in both their form and their habits to the Sand-wasps, among which indeed they are placed by Mr. Westwood and other writers. The females are wingless, of robust figure, have spinous legs fitted for digging, and are without the small simple eyes called ocelli. They are active insects and are found running on the ground in sandy places. The males are winged, and, as has been said above, are spicate at the tip of the abdomen. They have three ocelli, and their compound eyes are somewhat kidney-shaped, and larger than those of the female, which are round. In size the English species vary from 1-8th to 2-3rds of an inch, and the relative size of insects of the opposite sexes varies, the males being the larger in some species, and the females in others. The wings will be found in the table of wings of Hymenoptera.

The female of the largest European species, *Mutilla Europæa* (Pl. VII. fig. 2), can hardly have escaped the observation of the young entomologist, less because it is not very rare, than on account of its unusual appearance, which is that of a stout, hairy, wingless, red and black ant, of two-thirds of an inch in length. The male is smaller than the female, and somewhat varies in the distribution of its colours, but both are clothed with bands of pale glittering hairs, alternated with bands of scanty black down.

The habits of these solitary ants are as yet but little known, but it seems probable that they are parasitic in the nests of other insects, carnivorous, and predaceous. The female possesses a powerful sting.

CHAPTER XVI.

HYMENOPTERA.—ACULEATA.

THE second division of the predaceous stinging Hymenoptera, known as *Fossores* or diggers, consists of the Sand-wasps and Wood-wasps. From the true Wasps they are known by their fore-wings, which are not folded; from the Bees by their tarsi, of which the first joint is not wider than the following.

In general appearance some of them at first sight resemble the solitary species of true Wasps, others the Ichneumons, others, again, the gay yellow-banded parasitic Bees; but sufficient rules have already been given for distinguishing them from all of these insects. They vary much in colour and somewhat in form, some being black, others black and red, or black with creamy spots, others banded with bright yellow, and these latter are, like those of other banded and spotted insects, subject to much variation of marking. In form they are usually slender and wasp-like, with the abdomen in some attached by a decided stalk, while in others it approaches to being sessile. The abdomen is never laterally compressed as in some of the Ichneumonidæ.

The habits of these insects are interesting. The larvæ being insect-feeders, the parent forms a cell either (according to the species) in the ground, in the stalks of plants, willow, bramble, or rose, in old posts, &c., or in some tubular cavity or burrow which it finds ready made; here it deposits its eggs, and with them a store of insects to serve as food for the larvæ. Most species confine themselves to one kind of insect, but there are others that collect various kinds. Caterpillars, Spiders, Gnats, and other flies, Aphides, Beetles, Ants, and Bees, are all victims to one or another species. In some cases the prey is half killed, or reduced to torpidity, by being stung; in others it is stored quite alive, in others dead insects are laid up. Some observers have stated that there are species which are not content with laying up beforehand a store of food for their young, but continue to feed them at intervals. This Mr. Westwood doubts in the case of any solitary insect, though so well-known a habit with those which are social.

The land and wood wasps are divided into eight families:—1. Scoliidæ; 2. Sapygidæ; 3. Pompilidæ; 4. Sphegidæ; 5. Larridæ; 6. Nyssonidæ; 7. Crabronidæ; 8. Philanthidæ.

The first family, Scoliidæ, contains but two English species, of one genus, Tiphia. This genus may be known by the legs, which, in comparison with other Sand-wasps, are short and very thick, with wide, flat femora, and thickly-spined tibiæ. The wings have two sub-marginal cells. The antennæ are thick, and shorter than the thorax. *Tiphia femorata* is not rare. The female is entirely black, excepting parts of the legs, which are red. It is shiny, and scantily clothed with grey hairs; its length about ½ inch, or under. The male is considerably smaller, and bears a spine, curving upwards on the tip of the abdomen. This insect is very common on the cliffs at Lowestoft in Suffolk.

The second family, Sapygidæ, also contains but one English genus and two species. This, Sapyga, may be known by the kidney-shaped eyes and the presence of four submarginal cells in the wings. The other English genera of Fossores, with kidney-shaped eyes, differ in the venation of the wings. The antennæ are long, and somewhat club-shaped. The legs are slender and spineless; and thus the Sapyga is found making its cells either in burrows ready formed in the ground by other insects, or excavated by them in wood, or sometimes it makes use of small snail shells. *S. punctata*, the most common of the two species, is black, with small white markings on the head and thorax, the abdomen black and red, with white spots.

The third family, Pompilidæ, contains three genera. Pompilus (twenty species) is the principal; the others (Ceropales, with four, and Aporus, with two submarginal cells) containing together but three species, none of them common.

In Pompilus the wings have three submarginal cells, and the head is transverse; the antennæ are inserted in the middle of the face, and curled in the female. The hind legs are long; the abdomen is egg-shaped in the female, longer and more slender in the male, and attached by a very short stalk. The legs vary so much in different species, that the genus has been subdivided according to the presence or absence of hair fringes and spines, and this variety of structure affords an indication to variety of habit. These differences consist in the presence or absence of ciliæ on the tarsi of the fore-legs, and of spines (in double or single row, or irregularly placed), or of serrations in the tibiæ of the two other pairs. The hind legs are long throughout the genus. The colours are chiefly black; or black and red, or reddish brown, sometimes with white spots, wings usually somewhat dark. The various species of Pompilus are strong, fierce, and active insects, generally (though not without exception) making choice of Spiders, which they kill before storing them in their nests. They walk backwards with their prey in this way, carrying or dragging large Spiders for a very considerable distance. Some of the species burrow in hard seaside sandbanks, others in light sand; it has been said that some use ready-made burrows in wood.

Pompilus exaltatus (Pl. VII., fig. 3) is one of the commonest species. It is a bright and pretty insect, black and shining, with the exception of the abdomen, nearly two-thirds of which are red. The wings are darkish, with a pale spot near the tip, but this is sometimes absent in the females and usually so in the males. In this family the abdomen has a very short peduncle.

The fourth family, Sphegidæ, much resembles the former, but may be distinguished from it by the abdomen being set on a *long* stalk and the head on a small neck. There are four genera, all with three submarginal cells. This family contains but few species, of which *Ammophila Sabulosa* (Pl. VII. fig. 4) is the most conspicuous. This insect is sometimes nearly one inch long, and is black, with the central part of the abdomen red. It provides Caterpillars for the food of its young, in the storeroom at the end of the burrow, placing a Caterpillar first and laying one egg upon it, then adding three or four Caterpillars and carefully closing this burrow, it proceeds to form another.

The fifth family, Larridæ, is at once distinguished by the form of the mandible, which has a deep notch near the base on the outer side, and by the legs, which have one spine at the end of the tibiæ in the two first pairs, and two on the same place in the hind pair. The eyes in this family, except in one genus, Miscophus, approach closely in the female and become confluent in the male. Most of the species are rare. *Tachytes Pompiliformis* (Pl. VII. fig. 5) is a common insect by the seaside and in other sandy places. It may be known by the absence of the hind *stemmata*, the place of which is in this genus occupied by a tubercle. The insect is about $\frac{1}{3}$ inch long, black, excepting the fore half of the abdomen, which is of a red brown; the wings are darkish, and have one pointed marginal, and three submarginal cells. The fore tarsi are fringed with strong hairs, and the spines in the middle and hind pair are strong. It preys upon various insects, having been seen with Caterpillars, and by Mr. Smith, with a small kind of Grasshopper.

The sixth family, Nyssonidæ, brings us among the more wasp-like insects,* many of the species in this, and the two remaining families being banded or spotted on the abdomen, or abdomen and thorax with bright yellow. Many, however, exhibit the same colouring as the preceding families. In the Nyssonidæ the head is large, the mandibles are but slightly curved, not notched near the base, the antennæ are straight and threadlike, composed of short joints, the eyes ovate. The legs are somewhat spinous, the fore-legs have one comb-like spine on the tibiæ, and a corresponding notch opposite to this in the tarsus. The genera all have three submarginal cells in the fore-wings. The family contains five genera. The first, *Nysson*, may be known by the stemmata being (as is most common) in a triangle, and by the singular form of the abdomen, the second ventral segment abruptly forming an angle with the others; the colours

---

* In the preceding families but one yellow-banded species is found, *Sapyga clavicornis*.

are black and yellow, legs partially tinged with reddish brown, wings more or less darkened, *N. dimidiatus* is black and red, with white markings. The second genus, *Gorytes*, possesses in some species the latter peculiarity, but has the stemmata arranged in a curve. These genera contain some common species, black and yellow. The fifth genus, *Mellinus*, has the stemmata in a curve, and the petiole of the abdomen terminates in a knot. *Mellinus arvensis* (Pl. VII., fig. 6) is one of the most common of the Sandwasps. It is usually banded and marked as in the plate, but is subject to much variety in this particular. It is about ½ inch in length. The wings have a long pointed marginal cell, and four submarginal cells.* Of this insect Mr. Smith writes as follows: "Having frequently observed the habits of the *Mellinus arvensis*, and reared it from the larva state, a few observations are here recorded. When the parent insect has formed a burrow of the required length, and enlarged the extremity into a chamber of proper dimensions, she issues forth in search of the proper nutriment for her young. This consists of various dipterous insects; species of various genera are equally adapted to her purpose. *Muscidæ, Syrphidæ*, &c., are captured. It is amusing to see four or five females lie in wait upon a patch of cowdung until some luckless fly settles on it. When this happens, a cunning and gradual approach

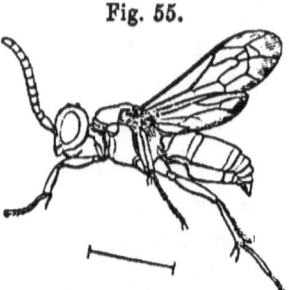

Fig. 55.

Profile Outline of Nysson Spinosus.

---

* The fourth, which reaches to the tip of the wing, is not shown in the plate, the nerve which bounds it falling short.

is made—a sudden attempt would not succeed: the fly is the insect of quickest flight, therefore a degree of artifice is necessary. This is managed by running past the victim slowly, and apparently in an unconscious manner, until the poor fly is caught unawares and carried off by the *Mellinus* to her burrow. The first fly being deposited, an egg is laid. The necessary number of flies are soon secured, and her task is completed; but sometimes she is interrupted by rainy weather, and it is some days before she can store up the quantity required. A larva found feeding became full fed in ten days; six flies were devoured, the heads, harder parts of the thorax, portions of the abdomen, and the legs being left untouched. The larva spins a tough, thin, brown silk cocoon, passes the winter and spring in the larva state, changes to the nymph on the approach of summer, and appears about the beginning of autumn in the perfect state."

The seventh family, Crabronidæ, much resembles the preceding in general character, but from it the three principal genera, Trypoxylon, Crabro, Oxybelus, are easily known from all other of the Fossores, by the presence of only one submarginal cell. The first genus, Trypoxylon, is distinguished by the eyes, which are deeply cut, or kidney-shaped, and the wings with one *tapering marginal*, and one submarginal cell. It contains three species, all common. *T. figulus* is to be found everywhere. It burrows in sandy banks, sometimes forming colonies, and provides its larvæ with spiders. It is a dusky black insect, long, with a long somewhat slender club-shaped abdomen, thickest at the end, contracted at the close of each segment, and with a slight hump near the base. The other two species are also black and similar in form. *T. claviceryum* and *T. attenuatum*, burrow in decaying wood; the latter also in rose and bramble sticks.

The second genus, Crabro, has large somewhat triangular eyes, rounded at the angles, and wings with one submarginal cell, which is truncated and has a fragment of nerve springing from the end.

The genus presents many varieties. In some species the abdomen is attached by a longer or shorter stalk, having a little hump at the termination, while in others it is almost sessile. In some the ocelli are arranged in a triangle, in others in a curved line. The legs are short, thick, and very spinous in most species,* and the males of some have the basal joint of the front tibiæ much dilated. In one species, *C. cribrarius*, the basal joint of the tarsus forms a broad thin plate, giving the limb a deformed appearance. The antennæ also are various in form. The colours in this genus are black, black and reddish brown, or black banded with bright yellow. In all but three of the species (of which there are thirty-six), the legs are partially of a bright yellow.

The different species form burrows in sandbanks, in wood more or less decaying, in brambles and rose sticks. One species, *C. luteipalpis* (one of those with no yellow on the legs), which burrows in the mortar of old brick walls, stores up the aphis as food; another, *C. brevis*, living in sandbanks, has been seen with a small species of beetle, but nearly all of which the habits are known, feed their young on various kinds of diptera. *Crabro vagus* (Pl. VIII., fig. 1) is one of the commonest species.

Of the next genus, Oxybelus, but one species, *O. uniglumis*, is very common. Mr. Smith describes it as

---

* This is generally the case with burrowing insects.

springing on its prey (two-winged flies) after the manner of a cat. The eyes in Oxybelus are ovate, the antennæ short, the legs thick, ciliated and spined, and the thorax has a sharp curved spine, near the base, which Crabro has not. The wings have but one submarginal cell, from which springs a short nerve as in Crabro. *O. uniglumis* is black with some white spots about the thorax, the abdomen banded and spotted with yellowish white, the legs black, reddish brown, and yellowish white.

In the genus Diodontus, the head is wider than the thorax, the abdomen attached by a very short stalk, eyes ovate, wings with two submarginal cells. The insects are small and black, with a little colouring about the legs. They prey upon aphides, and burrow in rose and bramble stems, or in sandbanks. One species, *D. minutus*, has colouring on the thorax, and yellow mandibles.

Pemphredon contains only one species. *P. lugubris* (Pl. VIII., fig. 2), an exceedingly common insect, is black, from $\frac{1}{3}$ to $\frac{1}{2}$ inch long, with a large head and a small glittering abdomen, which is attached to the thorax by a long and curved peduncle. The wings have two submarginal cells. It burrows in decaying wood, and has been observed by Mr. Smith to " settle on a rose tree, and scraping a number of aphides into a ball, fly off with it, carrying it in front of its anterior legs and under its head."

Mimesa equestris, a very pretty little insect, about $\frac{1}{3}$ inch long, is black, with the middle part of the small petiolated shining abdomen red. It seems not to be common except at Lowestoft. In the male the abdomen terminates in a spine, curved upwards. In this genus the submarginal cells are three.

The eighth and last family, Philanthidæ, consists of

but two genera, Philanthus and Cerceris. In both, the head is wider than the thorax, the tibia of the second pair of legs has but one spine at the end, and the fore-legs are strongly fringed with hairs on the tarsi. The fore-wings have three submarginal cells. In Philanthus the eyes are slightly cut, or inclined to kidney-shape, the legs are strong and spiny, the tarsi strongly fringed, the abdomen is ovate. *P. triangulum*, the only English species, is a beautiful insect more than $\frac{1}{2}$ inch long. The thorax is black, with creamy markings, and the face creamy. The abdomen is yellow with a black border narrowed in the middle to each segment, and a series of triangular black spots down the middle, decreasing in size towards the end. The legs are black and yellow. The male has a yellow line behind the eyes, and the abdomen is black with yellow bands, thinnest in their middle, and yellow on the two last joints. It feeds its young upon wild bees.

Cerceris may be distinguished by the decided constriction of each segment of the abdomen. This character occurs in Trypoxylon, and in a slighter degree in Philanthus. The antennæ are inclined to be clubshaped;

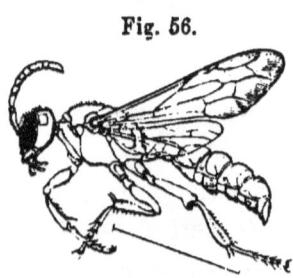

Fig. 56.

Profile of *Cerceris Arenaria*.

the legs strong, with strong spines, and with fringes on the fore-tarsi. The colours are black and yellow. This genus provides its young with beetles, amongst which are some of the hardest species; Mr. Smith, however, suggests that these, remaining in the damp ground for some days before the larvæ are ready for them, are softened by the time they are required as food.

CHAPTER XVII.

HYMENOPTERA.—ACULEATA.

THE true Wasps, solitary and social, form the third and last division of the Predaceous Hymenoptera, *Diploptera* (Διπλόω, *diploo* to double ; πτερὸν, *pteron* wing). As this name indicates, the Wasps are distinguished by the longitudinal folding, or doubling, of the fore-wing. The eyes of Wasps are kidney-shaped, the tongue is cleft and glandular at the tips; the first and second pairs of legs have one spine at the end of the tibiæ ; the hind pair has two spines. The claws are simple in the Social Wasps, cleft in the Solitary, and the wings of all have three submarginal cells.

The first family, the Solitary Wasps, or Eumenidæ, form two genera, Eumenes and Odynerus. To the first belongs only one British species, *E. coarctata* (Pl. VIII., fig. 3), which constructs upon the twigs of heath or other shrubs, a small round nest of mud in which it places a single egg, and a store of small caterpillars. This Wasp is about half an inch long, the male smaller. It may be distinguished from the Odyneri by the long pear-shaped stalk by which its abdomen is attached to the thorax. It is black with yellow spots and bands about the face, thorax, abdomen, and legs.

The second genus, Odynerus, contains twelve species of various habits. Some burrow in sandbanks, others in the pith of brambles, &c. : while others form their cells in any convenient receptacle which offers itself; Mr. Smith mentions a pistol-barrel, a piece of folded paper, and the hollow reeds in thatch, as having been chosen for this purpose by *O. quadratus*, a species which on other occasions burrows in old posts.

The young student will find no difficulty in distinguishing the Odyneri, the pear-shaped abdomen and stalk of the Eumenes sufficing to mark that genus, while the bifid claws of the Solitary Wasps divide them from the Social. The species are all black marked with yellow.

The cells of *Odynerus Antilope* (Pl. VIII., fig. 4) found by Mr. Smith in sandbanks, may very commonly be observed built up of mud, in the crevices formed by the perpendicular mouldings round doors, windows, &c., long mud tubes filling these hollows. The writer has seen the joints in a wooden summer-house filled by such cells from three to six or eight inches in length, and containing alternately a single Wasp-grub and six or seven emaciated green caterpillars. The little mothers appear to prefer a warm aspect for their young, frequently choosing a south wall, exposed to the full heat of the sun.

The second family, Vespidæ, contains the Social Wasps, of which there are in England seven species. They are divided into the Ground-wasps and the Tree-wasps, but their habits are not invariable, large nests of the Ground-wasps being occasionally found suspended in the roofs of houses and other such situations.

The Social Wasps, like the Social Ants and Bees, consist of males, females, and workers or small imperfect females, and their economy, though differing in some important details, is to a great extent similar. One point of difference is, that while the societies of Bees and Ants continue from year to year, those of the Wasps are strictly annual.

The foundation of a nest is laid in the spring by a solitary female, which having crept into some sheltered place at the approach of the winter, has survived its rigours, and now issues forth prepared to found a new city. Before the end of autumn this will have contained a population of many thousands. If a Ground-wasp (and we will take *Vespa Vulgaris* as the example), she commences her operations in some convenient cavity in the earth, it may be an old molehill, or a cavity under the roots of a tree. Here, of paper moulded of the gnawed fibres of wood, she constructs a small comb of a very few shallow cells, and, roofing it over, deposits an egg in each cell. She then proceeds to form more cells and lay more eggs; and, those first laid being speedily hatched, her labours in behalf of the young become unremitting. Not only does she feed them with the greatest care, but as they increase in size each little cell is again and again increased in depth. This forming of imperfect cells in the beginning points to a most curious economy of time. From the first eggs are hatched the larvæ of workers only, and it is evident that the increase and prosperity of the nest greatly depend on a speedy supply of labourers at this time. This the queen provides for by spending no more time in building than is absolutely necessary before she lays the first eggs, which she does as soon as the cells will contain them, trusting to her own unceasing activity to make up all deficiencies as occasion requires. These larvæ, then, she feeds and tends until the time of their first change. On emerging from the pupa state the young workers, within a few hours, set earnestly about assisting the foundress in her labours. They form fresh combs, they increase the size of the cavity in which the nest is placed, and, cutting up the original saucer-like covering of the nest, they use its material towards the construction of an elaborate roof of layer after layer of grey paper, the size of which increases with that of the nest itself. All this while, and indeed throughout her life, the female assists in these labours, not, as with the Ants, relinquishing such cares so soon as she is surrounded by a hundred little hands and feet willing and eager to undertake the whole labour of the hive, nor, as with the Bees, consenting to be installed in all the pomp and dignity of monarchy. The Wasp, on the contrary, having reared her brood of workers, proceeds to fill the new and refill the old cells with eggs which again are to produce workers only, and joins the first brood in the task of tending and feeding the second. This, however, is not all: the workers themselves begin to increase the population of the hive (although no males have as yet been hatched, these never appearing till towards the end of the season) and lay eggs which produce workers only, or, later, workers and males. The large, or perfect, females are always the progeny of the first mother or foundress of the nest, as, in the time which is approaching, these also will alone survive the winter, to be themselves the founders of new colonies.

When the colony has arrived at what may be called its perfect state, consisting of males, females, and neuters, the work proceeds more actively than ever. Living in perfect harmony, the many females now assist in the populating of the nest, sharing meanwhile the labours of the neuters; and the males, though they neither feed the young nor help in building, yet find themselves occupation in the way of " odd jobs about the house." Unlike the drones among the Bees, which seem to live only on sufferance, the male Wasps, acting as scavengers, undertakers, &c., are a welcome and useful portion of the community.

Throughout the summer, then, the varied labours of these citizens continue, the chief work being the care and feeding of the young. These, supplied at first with juices of fruits and such like tender fare, are presently promoted to an animal diet composed of insects or meat, half digested for them by their careful nurses; and this as they approach their full growth, is exchanged for the stronger nutriment afforded by these substances in almost their natural state.

There is little left to add to this history except the closing scene. It has been said that the societies of Wasps are strictly annual. Like all other Hymenopterous insects, Wasps are keenly sensitive to change of temperature, and the first few frosts are fatal to them.

What, then, is the lingering death in store for the young, hitherto so carefully fed and tended? Warmly sheltered in their little cells, it seems that they must survive their tender nurses, to die of gradual starvation, instead of by the quicker operation of the frost. But this is not the way in which such things are ordered. The nurses, for whom no labour has seemed too great, whose care for their young has up to this time been increasing, now suddenly seize upon them, and, tearing them from their cells, kill, without exception, every single grub, and scatter the bodies outside the desolated nest.

By this expedient, an expedient second only to that found in the marvellous system of prey, a quick and easy death is substituted for one of slow privation and suffering, and the parents and nurses die the most enviable of deaths, leaving none to miss them, and no work unfinished.

It is a well-known fact that the female insects in many orders are extremely tenacious of life until they have fulfilled their appointed work of continuing the race. Thus the life of a Moth or Butterfly, which under ordinary circumstances would terminate in a few months, may, if that be hindered, be prolonged to two or even three years. To this law it is perhaps owing that a few of the late hatched female Wasps survive the cold which destroys the rest of the community, and are thus ready at the return of spring to lay the foundation of a new nest. Let then the whole race of Wasp-haters bear this in mind. The single Wasp which trusts to the deceitful courtesy of one mild day in December or January to venture into our sight, will, before autumn, be the mother of some thirty thousand. She crawls forth half starved, half frozen, to claim from you perhaps the hundredth of a grain of one of your lumps of sugar. If you must murder Wasps, murder *her*, and fulfil the desire of a Nero—at one blow you have slain the thousands of a city. But when summer comes refrain from the useless cruelty of taking life after life from the joyous, busy little creatures whom you may kill by thousands without making the slightest perceptible difference in their numbers, although with every little victim one happy life has been quenched. If the preservation of fruit trees is the object in this random, useless warfare, the object will be better attained by placing more attractive food in the neighbourhood of the fruit. If their "nasty sting" is the objection to them, make but two calculations. First, inquire of half a dozen septuagenarian friends how often they have been stung in the course of their lives, and see if the average amount to more than one Wasp sting in thirty years. Secondly, reckoning how many millions of Wasps you may count upon as neighbours during those thirty years, calculate how much your chances of being stung are diminished by the number of those that you kill. If after this you still feel that your duty to yourself requires it, then by all means kill the next little nurse or mother that comes to see whether some of your breakfast would be nice for the little ones at home.

If the common saying that a good plum season is a season of many Wasps be true, we may find in it some comfort under their depredations.

It is impossible to enter here into the details of the architecture of the Wasp; suffice it to say that the nest spoken of above consists, when finished, of several large combs, placed horizontally one above the other, with the mouths of the cells downwards, and connected by strong pillars, or rather ligaments of paper, and roofed with a series of layers of grey paper. When, as sometimes happens, the Wasp builds her nest not in the ground but under the roof of an outhouse or loft, the roof is rather differently constructed, and looks like a loose tiling of small oyster-shells. The material with which the nests of the Tree-wasps are made is much tougher than that manufactured by the Ground-wasps, Mr. Smith observing that "the Tree-wasps may be considered as cardboard makers, and the Ground-wasps as paper makers." The cells and roofs of the latter are sometimes exceedingly fragile, the Wasp using, according to circumstances, decayed or sound wood, but even in this case preferring those parts which are worn by exposure. The oyster-shells forming the roof of a Wasp's nest, lately found in the roof of a dwelling-house, were beautified with zones of green, the little architects having made use of decayed wood coloured by the spores of P. æruginosa.

The species of Solitary Wasps are not always very easily distinguished, and would require a more minute description than space will allow to be given here. The females and workers of the Ground and Tree-wasps may, however, be distinguished by the colour of the first joint of the antennæ. In the former (Ground-wasps, *i.e.*, *V. vulgaris*, Pl. VIII., fig. 5, 5a, *V. Germanica, rufa*) this is black, and in the latter (Tree-wasps, *i.e.*, *V. arborea, sylvestris, Norvegica*, Pl. VIII., fig. 6, 6a, and *V. Crabro*) it is yellow in front, as in the males of all the species.

CHAPTER XVIII.

HYMENOPTERA.—ACULEATA.

From the predaceous Hymenoptera we now turn to the "Flower-lovers," or Bees. Familiar to all as are the common Hive Bee, and the great velvety Humble Bee, there are many species, little less common, which the young observer can hardly persuade himself to accept as Bees. Some are little black glossy creatures, hardly larger than the common Ant; others, a little larger, are glossy black and red; others have a metallic lustre; and others again, as the parasitic Nomada, are banded and spotted with black and yellow, yellow and red-brown, yielding in showiness of colouring to none of the Wasp or Sand-wasp tribes, and greatly resembling some of these in form and general appearance.*

It becomes necessary, therefore, to look for some character which shall distinguish the Bees from other insects resembling them in form or colouring. This is found in the peculiar form of the hind leg, already mentioned, page 187 (see fig. 52, and compare fig. 57, p. 222); the first joint of the tarsus in the Bees being a flattish oblong or long triangular plate, whilst in the Wasps, Sand-wasps, &c., this joint is cylindrical.

The purpose of this modification of form in the leg of the Bee is discovered by observing the use made of the limb by the larger number of species. The flattened

---

* See Plate IX.

tarsal-joint (fig. 57, 1, 2, 3 *e*), and the tibia (*d*) to which it is attached, are in many Bees densely clothed with hairs for the conveyance of pollen, whilst in the neuter Social Bees (both Hive and Humble) these joints are also naked on the outer side, flat or slightly concave, and fringed with hairs, thus forming a kind of basket for the reception of the pollen. The reader can hardly have failed to observe the flight homewards of Bees thus laden, their legs appearing enormously enlarged, and coloured red, white, and yellow, according to the colour of the pollen of such flowers as they have been visiting. Thus the mignonette-bed sends out a host of red-legged Bees, the same Bees issuing from the hollyhock are laden with white pollen, and others carry home a store of gold.

This flattened form of the tarsus, existing more or less in all Bees, does not however always indicate that each Bee is a pollen bearer, nor does the absence of its pollen-bearing accessories prove a Bee to be one which lays up no stores. Thus in the male or drone of the Hive Bee, which takes no part in the collection of provisions, the first tarsal joint is remarkably large and flat in proportion to the rest of the tarsus, but it is not hollowed and fringed on the outer side like that of the worker. In the parasitic Bees the flattening of the joint is observable though not conspicuous, and there are, as might be expected, no pollen-bearing appendages; while in others (some of the Solitary Bees) their place is supplied by a series of brushes under the abdomen, or by

Fig. 57.

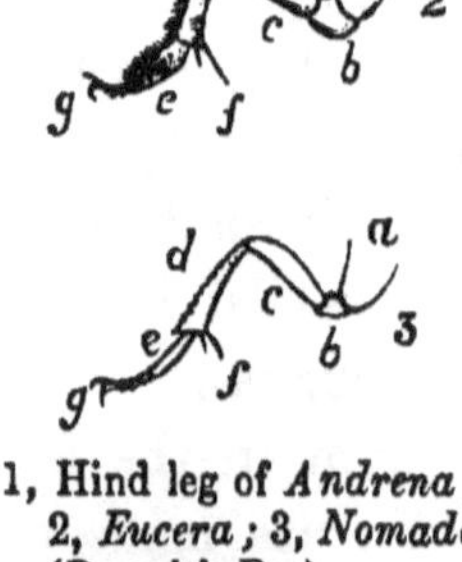

1, Hind leg of *Andrena;* 2, *Eucera;* 3, *Nomada* (Parasitic Bee).

pollen-baskets in the thighs and at the base of the thorax. Some Bees, known as builders and storers of provisions, are apparently without any contrivance of the kind, presenting one more of the countless paradoxes which arise on all sides in the investigation of nature.

The front legs of the Bees are furnished with a beautiful contrivance for the care and dressing of the antennæ. This is a comb-like moveable spur which grows at the end of the tibia, and closes down over a notch in the tarsus just deep enough to embrace the antenna. The Bees may be seen drawing their antennæ through these little notches again and again, cleansing them from dust and dirt, and even, when first emerged from the pupa, stripping off a membrane with which they are occasionally invested.

Setting aside for the moment all arrangement founded on structure, Bees may be distinguished as Solitary, Social, and Parasitic.

The Solitary Bees vary in their modes of life. Some make the tiny cells which are the cradles of their young in the hollow tubular stalks of plants, in snail shells, or in underground tunnels, and are in the strictest sense of the word solitary; while others, haunting in considerable numbers the same spot, form colonies, in which however each pair has its independent dwelling-place.

The Social Bees live either in republics or patriarchal (or rather matriarchal) communities, each household consisting (as with the Social Ants and Wasps) of one or more large perfect females, of smaller imperfect females or neuters, and later of males and the large females which are to produce their young in the following year.

The Parasitic or "Cuckoo" Bees make their dwelling

in the territories of their neighbours, whether Solitary or Social; each parasitic species being, however, limited in its choice of the species with which to take up its abode.

From this slight sketch it will be seen that, making allowance for the difference occasioned by variety of food, great resemblance exists in the economy of all the tribes of Aculeate Hymenoptera. Like the Solitary Wasps and Sand-wasps, the Solitary Bees cradle their young in nests or tunnels, placing with them a store of proper food. Like the Social Wasps and Ants, the Social Bees live in communities, and, by the help of neuter or imperfect individuals, provide for their young with a continuous and tender care. Like the Parasitic Wasps and Ants, the Parasitic Bees find shelter for their young in homes for which they have not worked; and though in the one case this is death to the rightful inhabitant, who falls a prey to his rapacious guest, and in the other guest and host often live together in perfect harmony, yet there is enough resemblance to mark the chain of relationship which binds these tribes together.*

The scientific division of Bees, based on their structure, depends chiefly on peculiarities in the tongue, legs, and wings.

In the first family, Andrenidæ, *the tongue is short* (as compared with the mentum, or chin) *and flat* (fig. 58, 1). It is broad, obtuse, and bi-lobed or notched (somewhat like that of the Wasp, but without glands at the tip) in the two first genera; in the six remaining genera it is pointed, and triangular or more or less lanceolate.

---

* In the following pages, the reader must be careful to distinguish between the *social* Bees, *i.e.* those living in communities formed of ♀, ☿, and ♂, and the *gregarious* or *colonizing* solitary Bees, of which many pairs burrow near each other.

In the second family, Apídæ (fig. 59, 2), *the tongue is long and cylindrical*, and makes a double fold under the mouth when in repose.

The Andrenidæ are all Solitary Bees, living either singly or in colonies; and therefore, there being no community, no public works are required, and no supply of extra workers is necessary. Each pair constructs its own nest, and only the two perfect sexes are known in these tribes.

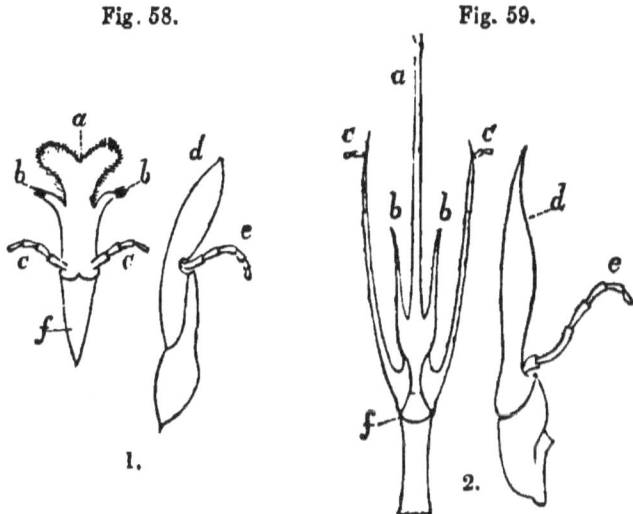

Fig. 58. Fig. 59.

* 1. Tongue, &c. of Colletes.   * 2. Tongue, &c. of Anthophora.
(*Andrenidæ.*)   (*Apidæ.*)

The two first genera of the Andrenidæ, Colletes and Prosopis, are easily distinguished from others by their obtuse bilobed tongues, and from each other by their appearance, the species of Prosopis being naked, while the Colletes are more or less hairy, the thorax being downy,

---

* *a*, Tongue; *b*, Paraglossæ; *c*. Labial palpi; *f*, Labium or mentum; *d*, Maxilla; *e*, Maxillary palpus. (Note.—*a*, *b*, *c*, *f*, are sometimes all included in the name of " Labium.")

and each ring of the abdomen fringed with grey or whitish hairs.

The Colletes are gregarious, and form large colonies. They burrow in the softer parts of walls or in sand-banks. The burrows, or tubular cells, are from eight to ten inches long, and the extremity is plastered inside with a thin coating of some substance the nature of which is not fully ascertained, though it is supposed to be a secretion of the insect. In substance it resembles fine goldbeater's skin, and it seems to be laid on in a soft or fluid state by the little trowel-like tongue.

The burrow being completed, an egg and a store of food composed of pollen and honey is deposited, and a cell is formed by sealing up the portion of tube filled with a flat wall of the same substance as that which lines the sides. A second and concave wall is then built and a second cell filled and furnished in the same way, till a series of thimble-shaped cells are formed. It is believed that the little mother is not always satisfied with a single burrow. The wings of Colletes have three submarginal cells. The species are from one-eighth to a quarter of an inch long.

The genus Prosopis differs from Colletes in its solitary habits as well as in appearance. Nearly naked, and without apparent means of collecting pollen, it has been supposed to be parasitic upon some industrious Bee; but while Mr. Smith's researches go to establish as fact that no species of the Andrenidæ is parasitic, it is now known that Prosopis forms cells (plastered, as by Colletes) in the stems of the bramble, rose, or other plants, which they render tubular by excavating the soft pith, and in which they make cells resembling those of the former genus.

The species of Prosopis are black, generally with

yellow or cream-coloured markings on the face (whence perhaps its name), on the thorax, and on the legs. They are from one-eighth to a quarter of an inch long; the wings have two submarginal cells.

The remaining six genera of short-tongued Bees have these organs pointed and more or less tapering.

The first genus, Sphecodes, consists of small Bees from one-eighth to half an inch in length, easily recognised by the absence of pollen-bearing organs and *sphex-like* colouring of their shining red and black bodies, the thorax being black and the abdomen brownish red (see Pl. IX., fig. 1, S. ruf.), sometimes tipped with black.

The Sphecodes are gregarious and frequently choose the same ground for their burrows as *Halictus*, the second genus of the sharp-tongued Andrenidæ; from this circumstance, and from the absence of visible means of carrying pollen they, like *Prosopis*, have been supposed to be parasitic. This, however, has been disproved by the observations of Mr. Smith, who not only saw the female Sphecodes at work upon her burrow, but found that the burrows entered by Sphecodes in a mixed colony were too small to admit the female of the Halictus.

The Sphecodes are black and red, and shining. The species, of which there are five, vary in size from *one-twelfth* (!) to half an inch. The submarginal cells are three in number.

The Halictus, gregarious, as its name indicates, forms burrows in the earth, in which several tubes open into one common entrance. These contain the eggs, and are stored with food as in other cases. This genus also contains some of the smallest Bees known, the ♂ of *H. minutissimus* is sometimes no more than one-eighth of an inch in length, *H. morio* (Pl. IX., fig. 2), an exceedingly abundant species, is one-sixth and over. This last is a beautiful little glossy black creature with somewhat of a metallic lustre; other species are more or less clothed or banded with white, grey, or golden hairs, while some are entirely black. The males have long slender bodies and long antennæ. It is worth noticing that in this and the preceding genus, the wing-hooks, instead of forming a series, arranged at regular, or regularly decreasing intervals, as in the generality of Bees (see fig. 24, p. 50), are interrupted and placed thus—

Fig. 60.

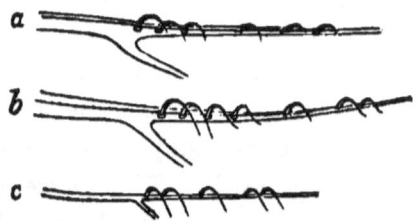

a. Hooks of posterior wing of *Sphecodes rufescens*.
b.   do.   do.   of *S. subquadratus*.
c.   do.   do.   of *Halictus morio*.

The examination of a large number of species might possibly prove this to afford a character useful in the distinction of species. The fore-wings have three submarginal cells.

The females of the genus *Andrena* may be recognised without difficulty by a beautiful little tuft of curled hairs on the underside of the trochanter of the hind-legs (see fig. 57, 1, p. 222). Being an instrument for conveying pollen, it is more conspicuous in the females than in the males, but these also have a tuft of hairs in the same position.

The determination of species is a much more difficult matter, and requires close examination. On the one hand, many species very closely resemble each other; while on the other, great variety in size, colouring, &c., occurs in individuals of the same species. A young and fresh coloured Bee can sometimes hardly be recognised when old and hoary, or when its golden glories are exchanged for the baldness of age. Besides this, the males are in some cases entirely unlike the females of their own species.

There is great variety of appearance among the Andrenæ, many of them are not unlike the common Hive Bee in size and colouring, &c., but they are generally rather more hairy, while others are sufficiently so to be mistaken by the tyro for small Humble Bees. Black, white, grey, and various shades of golden brown (Pl. IX., fig. 3, *A. fulva*), are the usual colours of the hairs, which are variously distributed, some species being thickly clothed all over, while others have the thorax furry, and the abdomen fringed with long hairs, or nearly naked. In one division the species are coloured like the Sphecodes, from which, however, the tufted trochanter easily distinguishes them. All the females are furnished with a thick pollen brush on the hind legs. In size the Andrenæ range from that of the Honey Bee down to one-third of its length. The submarginal cells are three in number. All the Andrenæ burrow in the ground, some burrowing alone, and others living in colonies, which very commonly include large numbers of the wasp-like parasitic Bee, *Nomada*, of which a description will be found in its place. The tunnels of the Andrenæ branch out in various directions, and are less elaborately finished than those of most other Bees.

The most noticeable insect remaining in the family of the Andrenidæ, is the *Dasypoda hirtipes*. It is slender and rather smaller than a neuter Hive Bee, and the female is rendered conspicuous by the enormous size of the pollen brushes, which clothe her legs throughout their length with a rich mass of golden fur. The body of the Bee is black and shining, the face and head are clothed with grey and black down, the thorax with black and golden. The abdomen has a little white down on the fore part, and its segments are fringed with white hairs. The males differ in the colour of their fur and in other particulars, and are less easy to recognise. The wings have only two submarginal cells.

The Dasypoda appears late, and forms burrows in sand-banks and other situations, choosing, according to Mr. Shuckard, a southern aspect, and situations where the ground is overgrown with shrubs. It is somewhat local, but very abundant where it occurs.

CHAPTER XIX.

HYMENOPTERA.—ACULEATA.

THE Apidæ form the second family of Bees containing the long-tongued Bees. In the habits of genera in this family we find more variety than in the Andrenidæ. Amongst them are found Bees which burrow in all imaginable situations, in the earth, in brick walls, in the stems of brambles and other such plants, in trees, in posts; some build thin cells, grain by grain, with sandstones, like the case of some Caddis Worms; others form them with a substance secreted in their own bodies; some line them with portions of leaf, others, like the Pelican, with down plucked from their own bodies. Some dwell in cities, each in his own home, others in families, hundreds living under one roof, while of the latter, some (the Hive Bees) live under strictly monarchical government, others (the Humble Bees) in a Republic.

And yet another mode of life remains. Some, brothers and sisters to these most ingenious architects, most tender nurses, most sober housekeepers—some, eschewing all the dull duties of life, light of form, bright in colouring, spend their little lives without care and without labour; appropriating to themselves the fruit of the toil of others, and, digging no tunnels, shaping no nests, collecting no stores, quietly provide for their young by depositing them in the well-stored nests of those who like such occupations, and thus "doing their duty by themselves and their children," afford a happy illustration of a favourite proverb—"Charity begins at home."

A likeness between the habits of these insects and those of the Cuckoo cannot fail to occur to every observer, and indeed to this they owe the name of *Cuckoo Bees*.* A closer examination by no means diminishes this resemblance; both are distinguished in their tribes by anatomical peculiarities in the mouth and feet. The bill of such of the Cuckoo family as are known to be parasitic is formed differently from that of species even in the same family which are nest-builders; and this, accompanying a certain form of foot in the parasitic birds, points to a relation between form and habit which at least exonerates the bird from the charge of voluntary idleness. In the same manner we find the jaw of the little *Nomada* very different from that of the burrowing

Fig. 61.

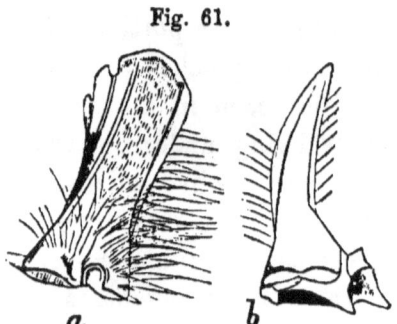

a. Jaw of *Bombus, Nest-digging Bee.*
b. Do. *Nomada, Parasitic Bee.*

or cell-making Bees, see fig. *b*, jaw of *Nomada;* and fig. *a*, jaw, or rather *spade* of the the Humble Bee, not

---

* Or *Cuculinæ*, a name applied to the whole sub-family, which includes five genera.

to mention the more elaborate instrument provided for the chief of Bee architects, the working Hive Bee.

The legs (fig. 57, 3, p. 222), and indeed the whole of the body of the Cuckoo Bees, are unprovided with the means of carrying pollen. It is true, as has already been noticed, that there are some other Bees, not parasitic, which also appear to be without these organs, but this is one of the difficulties which constantly encounter the student, and for which as yet no solution has been found.

But there is another point in which the Bee and the bird are alike—*in their care of their offspring.* The bird, much maligned in popular tradition, is yet an attentive and affectionate mother. Not confining her care for her young to finding for them the lodging which nature has denied her the power to construct, nor limiting her provision for them by the instinct which leads her to place her eggs invariably in the nests of such birds as feed their young with insects—she herself looks to their wants and brings them food. Lest this should be difficult to believe against the common prejudice, the testimony may be called of Dr. J. E. Gray, who for a considerable time watched a Cuckoo daily engaged in supplying her young with insects.

This the little Bee does not achieve, but there is reason to believe that she does what is equivalent. Mr. Smith relates that he has frequently captured species of these Bees *with masses of clay attached to their* posterior tibiæ, and his observations point to the conclusion that the cunning little creature seizes the moment when a stored cell is open and ready to receive the egg, that she then deposits her own, and *seals up the cell,* while the owner of the cell returning and finding either a sealed cell or a cell with an egg in it, deserts it, and commences a fresh receptacle for her own young. This, as Mr. Smith observes, is far more in accordance with the common course of nature than, as has been usually supposed, that the young Cuckoo Bee should eat the food provided for another and so starve that other; for, as he says, "nature I have never observed to be thus wasteful of animal life, such a proceeding is unnecessary and therefore unlikely."

This remark finds an illustration in the system of prey, the whole principle of which is adverse to such waste of life; this system both guarding animals from a lingering death of old age and starvation, and rendering their death immediately conducive to the life and enjoyment of others.

There are five genera of Cuckoo Bees. The first, *Nomada* (Pl. IX., fig. 4), are elegantly formed Bees, with nearly hairless bodies and wasp-like colouring, being banded with black, red-brown, and yellow, whence they are commonly called "Wasp Bees." They are parasitic on species both of Andrenidæ and Apidæ. *N. sexfasciata*, one of the largest of the species, may often be seen in numbers flying noiselessly over a colony of the long-horned Bee. It is half an inch in length.

Nomada contains twenty-four species, varying in size from one-sixth to the half of an inch.

The remaining genera, Epeolus, Cœlioxys, Stelis, and Melecta, contain some exceedingly pretty species, but are less showy than the Nomada. They are generally of a glossy black, with more or less white, creamy, or yellow down about the head and thorax, and sometimes with white stripes or bands of white down upon the abdomen. *Melecta luctuosa* is especially beautiful, of "shining jet spotted with snow white." Cœlioxys is remarkable from the acutely conical form of the abdomen, which is truncated in front.

These little creatures, of course, follow the Bees upon which they are parasitic, into whatever burrows they may form, and thus the student must be on his guard against taking for granted that the young Bee which he has carefully hatched from a cell in a bramble-stick, or has watched in its first emerging from one in an old post or wall, is necessarily the maker or the rightful owner of such cell.

This intrusive family of Cuculinæ, now described, has, according to its wont, pushed itself into a place where it had no right to be; and if a reader, more scientific than those for whom these pages are intended, should ask why the *Cuckoo Bees*, the *second family of the Apidæ*, have been allowed to come into the place of the *Andrenoides*, the *first family*, the only available answer is that the writer on Bees has had no better luck than the Bees themselves, and could not keep the Cuculinæ from coming where they had no business to come.

The sub-family of Andrenoides consists of but one genus, *Panurgus*, containing two species, and owes its proper position as first of the Apidæ to its resemblance to the Andrenoides, as indicated by its name (*i.e.*, Andrena-like). It does, in fact, appear to form a connecting link between the Andrenidæ and the Apidæ. In their manner of burrowing and storing their nests, in the possession of a pollen brush on the several joints of the legs, and other particulars, the Bees of this genus show a close affinity to the former, while the character of the folded tongue at once determines them to the latter family. They have, besides the pollen brushes, a shiny pollen basket on the outside of the thighs, and similar plates at the base of the thorax. The wings have two submarginal cells.

The third family, *Dasygastræ* (*i.e.*, the hairy-bellied Bees), is distinguished by a dense clothing of hairs forming a pollen brush on the underside of the abdomen of the female, and by the very large upper lip, which last, with the spines which terminate the abdomen in the males, suffices to mark these also.*

In all the genera except Ceratina, the submarginal cells are two in number.

This family contains some of the most interesting insects of the whole tribe; and perhaps a greater variety of architectural contrivances is found in it than in any other family.

Some species of Osmia, the first genus, have obtained the name of Mason Bees, from their use, in the formation of their cells, of a cement composed of small stones, grains of sand, &c., agglutinated together; but this habit is not confined to, nor indeed most remarkable in, this genus.

The reader will require no apology for the quotation of a passage in the interesting "Catalogue" of Mr. Smith upon this genus.

"If I were asked what genus of Bees would afford the most abundant materials for an essay on the diversity of instinct, I should, without hesitation, point out the genus Osmia. The most abundant species is O. bicornis (Pl. IX., fig. 5). Its economy is varied by circumstances. In hilly countries, or at the seaside, it chooses the sunny side of cliffs or sandy banks, in which to form its burrows; but in cultivated districts, particu-

---

* In Ceratina, the last genus, this pollen brush is wanting.

larly if the soil be clayey, it selects a decaying tree, preferring the stump of an old willow. It lays up a store of pollen and honey for the larvæ, which, when full grown, spin a tough dark brown cocoon, in which they remain, in the larva state, until the autumn, when the majority change to pupæ, and soon arrive at their perfect condition. Many, however, pass the winter in the larva state. In attempting to account for so remarkable a circumstance all must be conjecture, but it is not of unfrequent occurrence. This species also frequently makes its burrows in the mortar of old walls. Osmia leucomelana may be observed availing itself of a most admirable, and almost ready, adaptation for a burrow; it selects the dead branches of the common bramble; with little labour the parent Bee removes the pith, usually to the length of from five to six inches; at the end she deposits the requisite quantity of food, which she closes in with a substance resembling masticated leaves— evidently vegetable matter; she usually forms five or six cells in each bramble-stick. The Bee does not extract the whole of the pith, but alternately widens and contracts the diameter of the tube, each contraction marking the end of a cell. The egg is deposited on the food immediately before closing up the cell. It is white, oblong, and about the size and shape of a caraway seed; the larva is hatched in about eight days, and feeds about ten or twelve, when it is full grown; it then spins a thin silken covering, and remains in an inactive state until the following spring, when it undergoes its transformation, and appears usually in the month of June.

"Osmia hirta burrows in wood, seldom in any other material. The same habit will be observed in O. ænea;

but I have observed this Bee more than once constructing its burrow in the mortar of walls, and sometimes in hard sandbanks. O. aurulenta and O. bicolor are Bees which commonly burrow in banks; the latter being very abundant in some situations, forming colonies. But although it appears to be the usual habit of these species to construct tunnels in hard banks with great labour and untiring perseverance, still we find them at times exhibiting an amount of sagacity and a degree of knowledge that at once dispels the idea of their actions being the result of a mere blind instinct, impelling them in one undeviating course. A moment's consideration will suffice to call to mind many tunnels and tubes ready formed, which would appear to be admirably adapted to the purposes of the Bee. For instance, the straws of a thatch, and many reeds; and what could be more admirably adapted to their requirements than the tubes of many shells? So thinks the Bee. Osmia aurulenta and O. bicolor both select the shells of Helix hortensis and Helix nemoralis: the shells of these snails are, of course, very abundant, and lie half hidden beneath grass, mosses, and plants; the Bees, finding them in such situations, dispense with their accustomed labours, and take possession of the deserted shells. The number of cells varies according to the length of the whorl of the shell selected, the usual number being four, but in some instances they construct five or six, commencing at the end of the whorl; a suitable supply of honey and pollen is collected, an egg deposited, and a partition formed of abraded vegetable matter: the process is repeated until the requisite number is formed, when the whole is most carefully protected by closing up the entrance with small pellets of clay, sticks, and pebbles; these are firmly cemented together with some glutinous matter, and the Bee has finished her task.

"We will now observe the intelligence of the Bee under different circumstances. She has selected the adult shell of Helix aspersa; the whorl of this species is much larger in diameter than that of H. nemoralis, or H. hortensis, too wide, in fact, for a single cell. Our little architect, never at a loss, readily adapts it to her purpose by forming two cells side by side; and as she advances towards the entrance of the whorl it becomes too wide even for this contrivance. Here let us admire the ingenuity of the little creature; she constructs a couple of cells transversely. And this is the little animal which has been so blindly slandered as being a mere machine."

The antennæ of the males of Osmia are fringed with hair down one side.

The next genus, Megachile, is marked by the size of its jaws and by its two-jointed maxillary palpi. It contains the Leaf-cutters,—Bees which, forming their burrows in various situations, in the softer parts of old walls, in the ground, or in wooden posts, &c., partition them into a series of cells by means of circular pieces of leaf, so accurately fitted together as safely to confine the honey which is stored in them. Some species are not uncommon, and the reader may perhaps recall to his mind the frequent appearance of rose leaves (not petals) with circular holes in them, always cut *from the edge of the leaf*, and which have been operated upon by some of these Bees. It is easy to distinguish such leaves from those eaten by caterpillars, the clean-cut edge, the nearly perfect circle, and the approach to uniformity of size in these holes, marking the work of the Leaf-cutter

Bee. There are several bees parasitic on Megachile, Cœlioxys being the most frequently so.

The male in some species is easily recognised by the conspicuously dilated, fringed fore-legs, which appear like a mass of down; in some the last joint of the antennæ is flattened and widened; others are without these characteristics. The Leaf-cutting Bee described as lining its cells with the petals of the scarlet poppy, is not an English species.

Anthidium somewhat resembles Megachile in appearance; it forms cells in holes or tubes (not, as Mr. Smith believes, making these burrows, but using such as she finds ready), which she lines with down collected from the woolly leaves of certain plants. Mr. Kirby found such a nest in the keyhole of a door.

In this genus there is only one English species, A. manicatum, in which the male is larger than the female. It is a handsome black Bee, with yellow markings on the face, jaws, legs, &c., and a row of oval yellowish spots down the sides of the abdomen. The female is between one-third and half an inch, the male sometimes as much as two-thirds, an unusual circumstance among Bees, of which the female is nearly always the largest.

Chelostoma, the fourth genus of the Dasygastræ, contains only two species, the males of which are at once to be recognised by the curved abdomen, which is bent, or, it might almost be said, which is *curled* under them. *Chelostoma florisomnis*, the largest of the two species, being about one-third of an inch in length, is very commonly

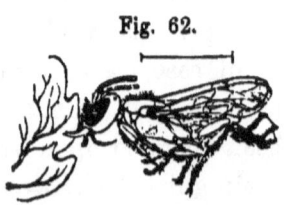

Fig. 62.

Profile of *Chelostoma florisomnis*.

to be found asleep in flowers, to which it is attached by the jaws (see fig. 62). In the female of this Bee the antennæ are very short.

Heriades closely resembles Chelostoma, and contains only one, and that a very rare species.

Ceratina is without the pollen-brush on the abdomen, but has long hairs growing sparsely on the legs;* it contains only one species certainly indigenous, and that is not common. It may be recognised by its glossy blue body, and the club-shaped form of its abdomen. It has, owing to the absence of polleniferous organs, been suspected of parasitism, but Mr. Smith has seen it in the act of excavating a dead bramble-stick. The submarginal cells are three in number.

The fourth sub-family is the *Scopulipedes* (or brush-legs). These Bees are furnished with a dense clothing of hairs on the hind-legs of the females.

The first genus, Eucera, contains but one species, the male of which is instantly to be recognised by the great length of the antennæ, which is equal to that of the body. As the Eucera is a colonizing Bee, it is easy to find males and females at the same time, at the end of May and in June. The antennæ of the females are not remarkable.

This Bee burrows six or eight inches deep in the earth, forming for its larvæ a curious oblong-oval brown cell of thin material, in which it places both larvæ and food. A colony of these Bees, settled on the green slope of a garden-plat or lawn, is a most enlivening addition to the pleasures of the garden. The males are busy,

---

* It is on the ground of general affinity that this Bee is placed among the Dasygastræ notwithstanding the absence of the abdominal hairs.

musical, good-natured, very playful, and their flight, when coursing one another over the ground occupied by their colony, is rapid and graceful. Additional interest is given by the intermingling of the bright little yellow-banded parasite Nomada sexfasciata, which appear to have free entrance to the burrows of the Eucera.

Another parasite, a two-winged fly, looking so like a Bee as easily to be mistaken for one, is also to be found with these Bees. The submarginal cells in the wings of Eucera are two in number.

Saropoda is distinguished by the labial palpi, which though really composed of four joints, form a straight bristle-like organ. In the other English Apidæ (see fig. 63, *e*), the terminal joint or joints are placed at an angle. It contains only one species, S. Bimaculata, a wood-burrower. The wings have three submarginal cells.

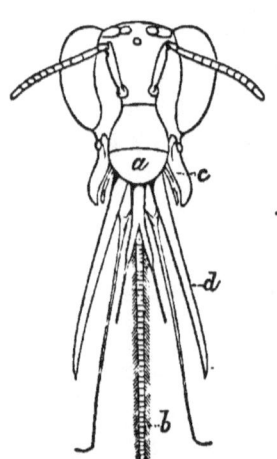

Head of Anthophora retusa.

Anthophora contains four species. They burrow in hard banks, old walls, &c., forming in them white-lined, elliptical egg-cells, which they provision and seal. A. Acervorum, a very common Bee, is nearly or quite two-thirds of an inch in length, black and hairy, with reddish-yellow hairs upon the legs in the female; in the male, the fore part of the abdomen is clothed with yellow hair, there is yellow about the face, and the legs, of which the middle pair are long, have fringes of white and of black hairs. The submarginal cells are three.

The last sub-family of Bees is the *Sociales*, containing the Humble Bees (Bombus), and the Hive Bees, besides a genus of large Cuckoo Bees (Apathus) which so nearly resemble the Humble Bees that, although supposed not to be nest-builders, they are naturally associated with them; all these have three submarginal cells. (Pl. IX., fig. 6. Bombus terrestris ♀.)

The Humble Bees are too well known to require description, and no others are likely to be mistaken for them, except the parasitic Apathus. Their large, heavy, handsomely-clothed bodies, their pleasant hum, and their busy habits, render them noticeable to the least observant, and they are almost universal favourites.

In habits they are social, or partially so; and, like the other social Hymenoptera, they consist of males, females, and neuters. In one respect their habits resemble those of the Ants rather than of the other Social Bees (the Hive Bees), namely, the existence in their community of several females at the same time.

The nest is founded in the spring by a single female, who either builds it of moss, grass, &c., or, if a ground Bee, chooses some ready-made hollow wherein to commence her work. Here she differs from the ground Wasp, who, under similar circumstances, never selects a cavity larger than is required for her present use, this being enlarged as it becomes necessary; while the Bee chooses one large enough to contain the nest when it shall have attained its full size. From the first-laid eggs (as in the case of the Wasps), only neuters are hatched. The nests are lined snugly with moss, grass, horsehair, or other suitable material, with the addition of a waxen cement. The young enclose themselves for their change in silk cocoons, whence, like such of the Ants as spin cocoons, they are aided in emerging by their watchful nurses.

There is a curious note in the "Zoologist" (3627), by Mr. E. Newman, who observed that Humble Bees—usually so good-tempered—become ill-tempered whilst awaiting the hatching of the neuters.

The Humble Bees differ greatly in size, even in the same species. The females are the largest, the males next in size, and the workers least, these being sometimes less than half the length of the female. There is sometimes great variety in the size of individuals even of the same gender. The colouring also varies greatly, and varies in the same individual at different periods, and the species are sometimes difficult to determine. The legs of the female Humble Bee have the usual pollen-basket, and the hind tarsi have a deep notch at the base, forming a tooth. This is wanting in the male.

The next genus, Apathus, consists of two sexes only, and is believed to be parasitic on Bombus. Little is as yet known, however, of the connexion subsisting between these insects. They closely resemble each other, but the females are easily to be distinguished by the legs, the tibia of which is concave on the outer side in the Apathus, while in Bombus it is hollowed into a pollen-basket, and the hind tarsus is without the deep-cut and prominent tooth so conspicuous in the Bombus. In Apathus, too, the abdomen is curved quite under at the tip, and great part of the upper side of the abdomen is almost bald and very glossy; the tongue has no paraglossæ.

It is remarkable that these parasites should so closely resemble the Bees on which they are parasitical, while others, as for instance the Nomada, are so strikingly different from Eucera, &c., upon which they live. Mr. Smith suggests as an explanation of this, that as the Solitary Bee would repel any intruder even of her own species, such resemblance would be useless, and therefore does not exist, while in the case of the Social Bees it might be of great service to the interloper. In the case of Solitary Bees, too, only one watchful little mother has to be supplanted, while the Humble Bee nest has many pairs of interested little eyes looking on at all that happens.

It seems, however, that the idea of any deception or evasion being necessary, or intended, is proved untenable by the harmony which exists between the solitary Bees themselves and their parasites, which goes to show that the Bee does *not* always repel intruders.

The last genus is Apis, which contains only one English species, but that the crowning wonder of the whole tribe—the Hive Bee.

So much has been written and said of the habits of this insect, that such space as Bees could claim in this work has been devoted to the description of less-known genera, and it is not proposed to enter into details of the life of the Hive Bee, which the reader may find in other volumes without number.

The Hive Bee is commonly quoted in connexion with the Instinct *versus* Reason question, but whilst the "instinct" of this little creature — as displayed in its admirable system of architecture, in its unvarying adherence to the established form of government, in its manner of collecting and of storing up food—is a favourite topic with those who allow only "instinct" and deny "reason" to the lower orders of animals, it must be admitted that, for beings without reason, the Bees sometimes act in a remarkably reasonable manner. Thus, some bees transported to Peru ("Zool." 7574), and providing a plentiful supply of honey in the first season, gradually slackened in their labour, diminishing the quantity stored up, until they nearly ceased to collect at all. Were the Bees indeed possessed of reason, this might be explained by their finding that flowers were to be had all the year round, and judging from thence that the labour of collecting stores of honey was superfluous.

In the same periodical from which this incident is borrowed, there is an account of a huge *colony* of Hive Bees which had taken up their abode in a blank attic-window of an old house in Yorkshire. A swarm of Bees had settled in the casement, and had been left undisturbed for some years, during which time swarm after swarm had yearly been thrown off and had settled in the same old mullioned window, and, when taken, the nest consisted of tier above tier of combs, several feet in height, and weighing, with the honey, 160 pounds. Here the common habit of the removal of the young swarms was not followed, there being accommodation for them near their old homes.

If the foregoing stories go to prove that the "instinct" of the Bee includes the power of comparing facts and drawing conclusions from them, the following anecdote (from Ireland) gives an example of this same "instinct," exercised in the form of memory and resentment.

"Some Beehives were placed on a stand of two deal planks, not fitting closely together. After a time, the crevice between the planks increasing, the Bees began to use it as a passage, and at last commenced building combs beneath the stand. The gardener, disapproving of this step, broke down some of the combs. The Bees, who had before been most quiet and harmless, became so angry at the destruction of their work, that they began a systematic war against all human beings, getting more and more wicked every day, till at last we could get nobody to work in the garden, or even to go into it, and we were obliged to pick all the vegetables for dinner ourselves, early in the morning, before the Bees were regularly astir. They must have had very little sleep, for early and late they were ready to fight, and we were stung before half-past six in the morning, and at past ten o'clock at night. They had scouts constantly on the watch, and when we went into the garden we were met by one Bee, who did not attack us at once, but went off for a reinforcement, and then came back to the assault. One of us got ten stings in one of these encounters.

"At last the Bees became so bold that they actually came round to the hall door, and we began to think that they must be exterminated, which, as they could not be smothered in the usual way (the most mischievous being those living under the stand), was at last effected by three or four men with spades and hot water.

"Our usual theory about Bees is, that if you let them alone, they will let you alone; but in this case, except for the original offence of breaking their combs, their attacks were quite unprovoked."*

The Hive Bee, whether male, female, or neuter, is distinguished not only from all other Bees, but from all

---

* It is said that there is in Mexico a species of stingless Honey-Bee which goes by the name of Angelito, or Little Angel. Somewhat of a contrast to our little Irish friends.

other Hymenopterous insects, by the absence of spurs at the end of the hind tibiæ. In the neuters the pollen-bearing hind tarsus has a peculiar and beautiful arrangement of the hairs, which form a series of regular transverse lines across the limb. The pollen-brush on the legs is entirely wanting in the Queen Bee—a circumstance which occurs in no other female of pollen-collecting genera.

As there is but one queen in a hive, it would be a mischievous act to capture her, even if by rare chance we should meet with her. It may therefore be useful to the young student to know that specimens of the female Hive Bee are to be found in June and July lying dead near the mouths of the hives, being the young queens sacrificed to that rule in the Bee monarchy which suffers no rival near the throne.

The substances or materials collected or produced by Bees are four in number—honey, bee-bread, wax, and propolis. Of the first of these it is needless to say much: the Bee collects it in a pure state from flowers, swallows it, carries it home in the *honey-bag*—a sort of first stomach, somewhat resembling that of cud-chewing quadrupeds—and then disgorging the greater part, either imparts it to other Bees in need of food, or stores it in the cells for future use.

The Bee-bread, or pollen-paste, is used chiefly in feeding the larvæ, and is composed of the pollen or dust of flowers, which has undergone the process of partial digestion by the worker. As with the honey, all which is not immediately wanted is laid up in store.

Wax, the third and most valuable material, is a secretion of the Bee itself, in whose body—a little living laboratory—part of its constituents are extracted from the honey, and combined with fatty and other matter, when it exudes in little scales from between the rings of the abdomen in the state of pure wax.

The fourth substance is "Propolis," a dark-coloured resin, which the Bee gathers from the buds and wounds of various trees. This is used in the hive for various purposes. It varnishes the combs and strengthens the edges of the cells, and is used on various emergencies, either to stop crevices, or occasionally to seal up and cover obnoxious objects which may have intruded into the hive.

In seasons when honey is scarce, Bees will eagerly feed on other sweet substances—honey-dew, the juice of fruit, or over-ripe or bruised gooseberries, &c., but it appears that from these things the wax cannot be elaborated.

The enemies of Bees are very numerous. Besides the Cuckoo-like parasites of their own tribe, the "Ruby-tails," and the Mutillæ, which follow the burrowing wild Bees into their homes, they are subject to the persecution of personal parasites, as the Stylops (see p. 101), not uncommon on Hive and other Bees, and the acari or mites, which abound to a conspicuous degree upon the Humble Bees. Some fall a prey in their larvahood to the Ichneumon tribes, and others are carried bodily away by other predaceous Hymenoptera. Earwigs also destroy enormous numbers of underground Bees. In their very hives, too, they are beset with enemies. A toad has been seen to sit at the mouth of a hive and devour the Bees one by one as they appeared. The larva of the Death's-Head Moth finds its way to the interior, where it feeds upon the honey stored there ; the larva of another genus of Moths—still worse—devours the waxen caskets which contain those stores; and Wasps have been known boldly to attack and rifle the hives in open war.

Besides all this, while Bees occasionally collect for their own use honey which is poisonous to man, many flowers are poisonous to themselves. The tulip is one of those, from the cup of which it is said that a Bee rarely escapes alive.

TABLE OF HYMENOPTERA.*

SECTION I.—TEREBRANTIA.

Abdomen of ♀ furnished with a sawing or boring ovipositor.
Hind legs.—The trochanter two-jointed.
Antennæ various in form and number of joints.

SUBSECTION I.—PHYTIPHAGA (Plant-eaters).

Abdomen attached to the thorax by its whole width.

I. Serrifera (Sawbearers).

Ovipositor of ♀ in form of a double saw.
Tibia of fore-leg furnished with two spines.
Tongue trifid.

Family 1. Tenthredinidæ (Sawflies).

II. Terebellifera (Borers, or Wood-eaters).

Ovipositor of ♀ in form of a strong borer.
Tibia of fore-leg furnished with one spine.
Tongue simple.

Family 2. Uroceridæ (Wood-borers).

SUBSECTION II.—ENTOMOPHAGA (Insect-eaters).

Abdomen attached to the thorax at one point, or by a small stalk.

---

* Some of the characters in this Table, used to facilitate reference, are merely empirical, and apply to British species only.

I. Spiculifera (Dart-bearers).

Ovipositor of ♀ dividing into separate bristles.
(Larvæ parasitic in bodies of living insects.)

Family 1. Cynipidæ.
2. Evaniidæ.
3. Ichneumonidæ.
4. Chalcididæ.
5. Proctotrupidæ.

II. Tubulifera (Tube-bearers).

Ovipositor sting-like, but without poison.
End of abdomen retractile, like a telescope.
(Larvæ parasitic in nests of insects.)

Family 2. Chrysididæ (Ruby-tails).

SECTION II.—ACULEATA (Stinging Insects).

Abdomen of ♀ furnished with a venomous sting.
Hind-legs.—The trochanter one-jointed.
Antennæ twelve-jointed in ♀, thirteen-jointed in ♂

SUBSECTION I.—PRÆDONES (Predatory Insects).

Hind-leg.—Basal joint of tarsus not dilated.

I. Heterogyna (Different females), *Ants*.

Fore-wings not folded.

* ♀ with or without wings, and ♀ and ♂ with one or two nodes on base of abdomen.
(Habits social.)

  a. 1 *node.*

Family 1. Formicidæ.
2. Poneridæ.

  b. 2 *nodes.*

3. Myrmecidæ.

* * ♀ without wings, and ♂ furnished with spikes at end of abdomen.
(Habits solitary.)

4. Mutillidæ.

TABLE OF HYMENOPTERA. 253

II. Fossores (Diggers), *Sand and Wood-wasps.*

Fore-wings not folded.

♀ and ♂ always winged.

♂ no spikes at end of abdomen.

♀ and ♂ no nodes on base of abdomen.

Family 5. Scoliidæ.
6. Sapygidæ.
7. Pompilidæ.
8. Sphegidæ.
9. Larridæ.
10. Nyssonidæ.
11. Crabronidæ.
12. Philanthidæ.

III. Diploptera (Doubled-wings), *True Wasps.*

Fore-wings folded.

Family 13. Eumenidæ (Solitary Wasps).
14. Vespidæ (Social Wasps).

SUBSECTION II.—ANTHOPHILA (Flower Lovers), *Bees.*

Hind leg.—Basal joint of tarsus dilated.

I. Short-tongued Bees.

Tongue shorter than the mentum.

Labial palpi of four nearly equal joints.

Fam. 1.—Andrenidæ.

    * Tongue broad, more or less cleft.

      Sub-fam. 1.—Obtusilingues.

    * * Tongue sharp-pointed.

      Sub-fam. 2.—Acutilingues.

II. Long-tongued Bees.

Tongue longer than the mentum.

Labial palpi of four joints, of which the basal or the second exceeds the two terminal in length.

254   INSECTS.

Family 2. Apidæ.
- * Hind legs hairy, from coxa to tarsus
  Sub-fam. 1. Andrenoides.
- ** Legs and abdomen not hairy.
  Sub-fam. 2. Cuculinæ*
- *** Underside of abdomen very hairy.
  Sub-fam. 3. Dasygastræ.
- **** Tibia and tarsus of hind-leg very hairy.
  Sub-fam. 4. Scopulipedes.
- ***** Males, females, and neuters living in communities.
  Sub-fam. 5. Sociales.

N.B.—Terebrantia is here arranged as by Mr. Westwood, and Aculeata as by Mr. Smith, in the Brit. Mus. Catalogue, slightly altered to make the terms of division agree with those of Terebrantia.

Fig. 64.

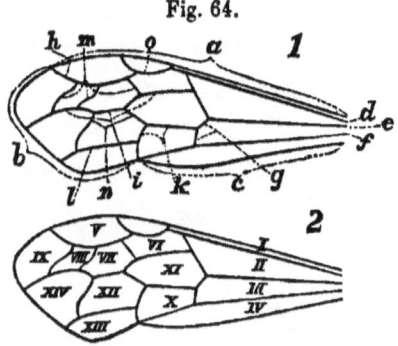

1. Nervures:—*a*, Costal; *d*, Post-costal; *e*, Externo-medial; *f*, Anal; *g, m*, Transverso-medial; *h*, Radial; *i*, Cubital; *k*, Discoidal: *l*, Sub-discoidal; *n*, Recurrent; *b*, Apical; *c*, Posterior margin; *o*, Stigma.
2. Cells:—I., Costal; II., Externo-medial; III., Interno-medial; IV., Anal; V., Marginal; VI., 1st; VII., 2nd; VIII., 3rd; IX., 4th. Submarginal: X., 1st; XI., 2nd; XII., 3rd. Discoidal: XIII., 1st; XIV., 2nd Apical.

---

* The naked-legged genus Ceratina is in Scopulipedes.

"There is a difference between a grub and a butterfly,
And yet your butterfly was once a grub."—*Coriolanus.*

CHAPTER XX.

ORDER IX.—LEPIDOPTERA.

THE Order LEPIDOPTERA is easily distinguished. It contains the Butterflies and Moths only, and excepting some species with clear and nearly naked membranous wings, and the wingless females of certain other species, which might for a moment perplex a beginner, excepting also a small number which, approaching the Trichoptera in character are a difficulty to the more advanced, Moths and Butterflies are recognised by even the most unobservant. More than this, not the appearance and names only, but even something of the history of these insects is very generally known, and observation of their caterpillars and chrysalids, or the beautiful sight of the young Moth emerging from the chrysalis, is often the beginning of a taste for the study of insects.

Their metamorphoses have already been noticed in

Chap. IV.; peculiarities in the larvæ and pupæ of some of the group will be entered into when these groups have been noticed.

The wings in Lepidoptera are four. They are large, and entirely, or in some cases only partially, covered by minutes scales arranged like the tiles of a house.* In some species the wings are furnished with a sort of spring, consisting of a strong curved bristle on the base of the hind-wing, which plays, during flight, in a socket or semi-loop, formed either of a ridge in the membrane, or of a tuft of hairs on the fore-wing. A curious epaulette-like appendage, called tegula, thickly clothed with hairs, of triangular form and sometimes of a large size, occurs at the insertion of the fore-wing.

The organs of sight consist of a pair of compound eyes, and frequently of additional simple eyes or ocelli.

The legs are hairy and spurred, and furnished with two claws of various forms. In some Butterflies the fore-legs are wanting.

Familiar as these insects are, and popular as they are among young collectors and students, they are rendered peculiarly difficult to treat in a very small space by the immense number of their families, genera, and species; and the absence of those marked differences of economy in the several families which, corresponding with marked differences of form, give so great an interest to other large orders of insects, as, for instance, to the order Hymenoptera.

---

* These scales, which, with other skin-appendages, as the scales of fishes, feathers of birds, &c., are somewhat of the nature of hairs, are, like them, rooted in the skin, and in some species are fixed there with great firmness by a club-like enlargement of their stalk at its insertion.

LEPIDOPTERA. 257

The numerous variations which occur in the habits of these insects, in the material of their food, and the situation or construction of their dwellings, almost exclusively concern the larvæ, and little or no indication of them is to be found in the imago.

Of the Butterflies only twenty-eight genera, containing sixty-six species, are known in England, and these are arranged in five families. Of the Moths, the known species of which are constantly increasing in number, there are more than one hundred families, consisting of between four and five hundred genera: these containing nearly two thousand species. The one hundred families are arranged in nine large groups.*

It will be easily seen that to describe the families (not to mention the genera) of the Moths, would be to reduce the following pages to little more than a mere table: the utmost, therefore, that will be attempted as regards arrangement, is to enable the reader to determine first whether an insect be a Butterfly or a Moth. If the former, to which of the five *families*, if the latter, to which of the nine *groups of families*, it belongs.†

First, then, with regard to distinguishing Butterflies from Moths. It is not at all uncommon to find an

---

* Minor subdivisions, as of sub-families, &c., are also in use, but it is not necessary to speak more of these here.

† In Mr. Stainton's volume on "The British Butterflies and Moths" (part of the present series), the reader will find an introduction to all the families of Moths, and to a large number of the more conspicuous species of both Moths and Butterflies. In his "Manual of British Butterflies and Moths" (2 vols.) the species also are described. The student who intends to make real progress in the knowledge of this tribe of insects is strongly advised to possess himself of the latter work. It has been largely drawn from in the following pages.

S

insect pronounced a Butterfly or a Moth because its colouring is bright or dull. Thus the dull-coloured brown Wood Butterflies are often supposed to be Moths, while the showy Tiger-Moth, with its rich brown and cream-coloured fore-wings, and hind-wings of bright scarlet and blue-black, the beautiful green and red Burnet Moth, and the Peacock-eyed Sphinxes, are called Butterflies. Sometimes, too, size is supposed to settle the question; and, on this account, the smaller Butterflies are called Moths, and the large Moths, Butterflies. This, like the appeal to colour, is quite erroneous. As to colour, nocturnal insects of all kinds are usually more soberly coloured than diurnal, and, the Butterflies being diurnal, while the larger number of species of Moths are night-fliers, the former are generally more conspicuously marked and coloured than the latter; but the white and brown Butterflies, and the numerous gaily-coloured Moths, make any rule, even in this matter, impossible. As to size, the range is much greater in the Moths, our largest Moth, the Death's-Head, sometimes measuring five inches from tip to tip of the expanded wings, while some of the minute leaf-mining Moths are smaller than the common little green Rose-Aphis. The largest Butterfly, on the other hand (the Swallowtail) seldom exceeds four inches, and the smallest, a little blue Butterfly, measures from about three-quarters to one inch.

Neither size, then, nor colour, will guide us in distinguishing between Moths and Butterflies.

Moths and Butterflies differ—first, in the form of their antennæ; secondly, in the position and folding of the wings when at rest; thirdly, very generally in the character of the caterpillar and chrysalis.

First, the antennæ. The two sections of Lepidoptera

are RHOPALOCERA ('Ρόπαλον, *Rhopalon*, club; κέρας, *keras*, horn), containing the Butterflies; HETEROCERA ("Ετερος, *heteros*, different; κέρας, *keras*, horn), containing the Moths.

The section RHOPALOCERA, the club-horns, contains the Butterflies. The antennæ of these are largest at the tip. In most they are very fine and hair-like, with an abrupt enlargement or knob at the tip. In others, as the Blue Butterflies, the enlargement is rather more gradual and club-like.*

In the HETEROCERA (the Moths), the horns are sometimes thickened about the middle or towards the tip, sometimes are like slender hairs, sometimes are branched and appear like exquisite feathers, but are always more or less tapering, being smaller at the tip than immediately below it.

Another difference between the Butterflies and Moths is, that in Butterflies the wings are never folded nor laid one over the other. In repose they are generally raised above the body and placed against each other, displaying only the under surfaces.

Of the Moths, on the contrary, while some repose with the wings expanded, the greater part fold the hindwings and lay the fore-wings down over both them and the body. The Butterfly-like Currant Moth, and some of its relations, rest with the wings raised Butterfly-fashion; but these having tapering and sometimes feathered antennæ, may by this be known as Moths. Some Butterflies of the family Hesperidæ (a family which seems to be in many respects a link between the

---

* There are exceptions to this rule among foreign Butterflies, some having tapering, hair-like, or somewhat flat-knobbed antennæ.

Butterflies and Moths), carry the fore-wings erect and the hind-wings horizontally when in repose.

The spring and socket (mentioned p. 256) is generally found in Moths and never in Butterflies. Again, the hind-legs of Butterflies have two pairs of spurs on the tibiæ (excepting in the Hesperidæ), while the Moths have only one pair.

The Butterflies are divided into five families—
1. Papilionidæ.
2. Nymphalidæ.
3. Erycinidæ.
4. Lycænidæ.
5. Hesperidæ.

1. Papilionidæ. This family includes (with one exception)* all the white, yellow, and greenish-white or yellow Butterflies, with and without black markings. The only approach to bright colouring in English species of this family is found in *Colias Edusa* (the clouded yellow), which is rich black and yellow, and in *Anthocharis Cardamine* (orange tip), of which the male has a patch of bright orange on the fore-wings, wanting in the female, and beautiful green markings on the under-side of the hind-wings. The large swallowtail—cream-coloured, with black markings—has also a brick-red spot on the hind-wings, almost the only instance of anything but black, white, yellow, and greenish when occurring in English species of this family.

The Brimstone or Sulphur Butterfly (Pl. X., fig. 1) is one of the most beautiful of these. The form is singularly elegant, from the varied curves in the outline of the pointed wings. The colour is delicate and beautiful,

---

* This is the Marbled White—the first species in the next family.

being true sulphur in the male, while the female is somewhat whiter.

Individuals are common of a tender greenish hue, which, combined with the angular form of the wings, gives a remarkably leaf-like appearance to the insect. The time of its appearance adds another charm, the Sulphur Butterfly being usually the first awakened from its winter sleep, while it is also one of the latest to remain with us, as if unwilling to give up the hope that "summer liveth still."

Those seen in the Autumn months are the lately-hatched individuals, the earlier visitors being such as have lain dormant through the winter.

The common large and small White Butterflies, the delicate and somewhat transparent "Black-veined White," the "Bath White," with its patches of black, and its greenish under-side, are all included in this family, which contains eleven species.

2. Nymphalidæ. The most striking peculiarity in this family is that all the species (in both sexes) have only four legs; the front pair being undeveloped. This distinguishes the Nymphalidæ from all other Butterflies but one—namely, the male of Nemeobius lucina, the only British Butterfly in the 3rd family.

In the Nymphalidæ the colours are generally dark, or rich and sometimes beautifully variegated; dark brown, rich tawny-brown, orange, black, with brilliant markings of scarlet, blue, and white, being all found here. One exception to them is found in a black and cream-coloured Butterfly, the "Marbled White," which in this assimilates with the Papilionidæ. The absence of the front pair of legs, however, at once marks it as belonging to the present family, and some ring-like spots, with white centres, are found on the underside of the hind-wings. This marking is very common in this family, but not in Papilionidæ, although the Clouded Yellow, Colias, in Papilionidæ has a ringed spot of somewhat silvery surface on the underside of the hind-wing.

The family Nymphalidæ contains, besides the Marbled White already named, all the brown wood and meadow Butterflies; sober-coloured insects whose chief charm is their very commonness, which associates them in our memories with the woodland scenes and sunny days in which they are seldom wanting.

There are several genera of these, and on the wings of all the species several small dark spots occur which have minute white centres, and are placed sometimes in a pale or tawny ring or patch. The common "Meadow Brown" (Pl. X., fig. 2), and the common little tawny "Small Heath," are examples of this. Amongst the more richly-coloured of the Nymphalidæ, we come to the White Admiral, and the Purple Emperor, both of a brownish black with broad white markings. The male of the Purple Emperor has also a deep purple lustre, and some ringed spots on each hind-wing.

In the Painted Lady, "painted" with a delicate mixture of rich dark brown with pale orange, tawny, and white, ring-like marks are conspicuous on the under side.

But it is in the genus Vanessa that the ringed or eye-like spot attains its highest glory in the Peacock Butterfly, which, with its gorgeous peacock-eyed wings, is perhaps the most *conspicuously* beautiful among British insects. In this the under side is of very dark rich brown—made the richer (as by "stippling") from being covered with a minute and obscure pattern of a darker colour. The upper side is too well known to require any description.

All the Vanessas, which include the Peacock, Tortoiseshells, Red Admiral, &c., and which, except the latter, are somewhat sober in the colouring of their under side, are remarkable for the ragged outline which they display when closed, but which the patterns and colours of their plumage render less conspicuous when expanded, and still more ragged than any of these is one in the next genus, Grapta, a rather scarce Butterfly, which a young entomologist might almost pass by, when at rest, as a torn and spoiled specimen. Above it is deep yellow, brown, and black; below, dark and dusky, and marked in the middle of the hind-wing with a c-shaped white spot, by which it may be recognised.

If among the Tortoiseshell and Peacock Butterflies a beautifully marked and coloured upper surface is often contrasted with a dark and dingy under side, in the Fritillaries which succeed them we find several species decorated underneath in an exquisite manner. In these insects the jagged and irregular outline disappears— the upper side is a rich tawny colour, distinctly lined and spotted with black: the under side of the wings is studded with spots (in one species with bands) of burnished silver. The *silvery* (*not white*) appearance of these spots is very remarkable, the effect being exactly that of the polished metal. A similar plumage is found in some minute moths, whilst in others appears a surface of true golden or brassy lustre. These silvered Butterflies are of the genus Argynnis; other Fritillaries, belonging to the genus Melitœa, in the same family, resemble them in their general colouring, but are without these spots.

We now come to the third family, Erycinidæ. This contains but one British species, Nemeobius Lucina, the male of which resembles the Nymphalidæ in the non-development of the fore-legs, while the female is six-legged like other butterflies. This, though distinct from the Fritillaries in the last family, resembles them in the colouring, and is commonly known by the same English name, *Fritillary*. It has no silvery spots beneath.

The fourth family, Lycœnidæ, contains those beautiful little blue butterflies (Pl. X., f. 3), which, haunting the same chalky districts that are the natural home of the blue harebell, are so often to be seen hovering over its delicate blossoms—so near, indeed, do these two little creatures approach in hue, in size, in fragility, in grace and beauty, that it needs little more than the languid dreaminess of an idle hour in a warm spring morning to see the flowers take wing, and to hear them whispering the secret of their delight to their less aspiring sisters, yet clinging to the slender stems which hold them to the earth. Besides these the family contains the brilliant little Copper Butterflies of the same size, and often to be found in their company, their dark, rich sparkling colour forming a beautiful contrast with the delicate hues of the little blues. The brown and orange, or purple Hair Streaks (Thecla) are also of this family.

The species of Thecla, may, all but one, be recognised by two small tails on each hind wing. The colours are brown with orange spots, or, in the Purple Hair Streak, brown, with a rich purple tinge. The one tailless species, the Green Hair Streak, is brown, without the orange spots, and may be recognised by its green underside.

The species of Blue Butterflies vary in hue, and the

females of some are brown, with or without a purple lustre. Some species of the genus are brown in both sexes, some rich brown with orange spots. The under side is pale blue, grey, pale greyish, or fawn-brown, covered or margined with black spots, some of which are enclosed in white, or whitish rings, sometimes with red spots. It is from the occurrence of these conspicuous spots that the genus Polyommatus (many-eyed) receives its name.

The Copper Butterflies, Chrysophanus, never have rings round the spots on their hind-wings. *C. Phleas*, the small Copper, is common, and is very frequently found in company with the Blue Butterflies, with the delicate hues of which its rich and burnished wings are in beautiful contrast, rendered more striking by the similarity of size and general form. This little Butterfly, like the Thecla, seems to have earned for itself the character of being quarrelsome. A curious variety has been found in which the copper was on both wings exchanged for pure white.

The fifth and last family, Hesperidæ, appears in several respects to be nearly related to the Moths, the body being thicker in proportion than is usual in Butterflies, the legs having, like the Moths, only one pair of spines, the fore-wings (in some species) being erect during repose, while the hind-wings remain in a horizontal position, and the antennæ in two species being slightly hooked at the tip, in a manner resembling that of some Moths. Besides this, the larva spins for its change a thin cocoon, a habit unusual among Moths, as will hereafter be shown. In this family the colours are chiefly rich brown and tawny, or yellowish. One pretty little species, Thymele Alveolus, is nearly black, with a greenish hue, and marked with numerous angular cream-coloured spots. The Butterflies in this family are remarkable for their short, abrupt flight, whence they derive their common name of Skippers.

The flight of Butterflies varies greatly in different families and genera. Those with the greatest power of flight are found among the richly-coloured species in the second family, Nymphalidæ, the Tortoiseshells, Peacock, Red Admiral, &c., one, the Purple Emperor, exceeding all others in this, soaring sometimes completely out of sight. In the same family, among the Brown Butterflies, are found some also of the weakest fliers, with a habit of keeping near the ground.

The Small White Butterfly, Pieris Rapæ (in the first family), though not reckoned among the strong fliers, distinguished itself about five-and-twenty years ago by flying from France to England in such countless swarms, that accounts of the time speak of the sun being completely hidden from vessels in the Channel, during a progress of several hundred yards, by the clouds of insects. It seems likely that they may have received the assistance of an aerial current on their journey, a strong west wind having arisen shortly after their arrival in England.

This *swarming* of certain species of insects in a particular year is a phenomenon which occurs in nearly all the orders, and is one of the problems in natural history as yet unsolved by observation. In some cases countless myriads make their appearance, as in the case of the Turnipfly recorded by Mr. Smith (p. 159), or the recurring instances of such swarms of Ladybirds. In others, insects more or less rare in some years, are comparatively abundant in others. This has been especially noticed of the "Clouded Yellow" Butterfly

(Colias Edusa), which, as Mr. Newman notes, is usually visible, sometimes exceedingly plentiful, every four years. The Camberwell Beauty, generally exceedingly rare, has been seen to come afield in a " flock," as have others of its more brilliant comrades, the Peacock, &c. The Leopard Moth, an insect so rare as to be formerly sold at a guinea a specimen, abounded about ten years ago to such a degree, that nearly all the young trees in Euston Square, and the copse plantations of ash in many parts of the country were destroyed by its pith-eating larvæ.

This may be partly accounted for by the power possessed by some Lepidoptera, of remaining long in the pupa state. Sphinx Ligustri has been known to remain in the Chrysalis for three years, while Mr. John Sircow records the coming out of a Moth after *six years* of incarceration. It may be supposed that a particularly favourable season brings to perfection the insects of many preceding years.

Mr. Douglas notes that weather has an effect upon the hatching of Lepidoptera, a warm rain after drought being favourable to this process. In some cases a great abundance of particular Butterflies has been accounted for by their having lived through the winter, thus adding the numbers of one year to those of another.

CHAPTER XXI.

LEPIDOPTERA—(*continued*).

THE families of Moths (Heterocera) number, as has been already said, about 100. These are formed into nine groups :—

1. Sphingina.
2. Bombycina.
3. Noctuina.
4. Geometrina.
5. Pyralidina.
6. Tortricina.
7. Tineina.
8. Pterophorina.
9. Alucitina.

The first group is readily distinguished by the *spindle-shaped* antennæ—antennæ, that is, which are thick in the middle, and taper towards the point and the base. They approach more nearly to the clavate antennæ of some Butterflies, than do those of any other group of Moths. And, indeed, similar antennæ are found in some foreign species in the last family of Butterflies—Hesperidæ.

The Sphingina have, by some authors, been considered not as Moths, but as forming an independent tribe, between, and equal to, the great tribes of Butterflies and

Moths. They are now, however, ranked as one group of the Moths.

This group contains, amongst others, the Sphinx Family,* many of which, though not among our commonest Moths, are well known, on account of the attention they attract when they do appear. Their bodies are large; most species have pointed and elegantly-formed wings, and some are rendered further conspicuous by the beauty of their colouring, and the eye-like spots on their wings. The brown and rose-coloured Privet Hawk-Moth, the Eyed Hawk-Moth, with rosy brown and bluish-eyed wings, and the small Elephant Hawk-Moth (Pl. X., f. 4) belong to the Sphinx family in this group. The Convolvulus, Privet, and Firtree Hawk-Moths are remarkable for their long tongues,—longer sometimes than the whole body, and able to reach into the honeyed depths of the longest flower-tubes. The singular and handsome Death's-head belongs to this family, and is as remarkable for shortness, as the true Sphinges for length of tongue— (Even the Sphinx loquacious as compared with Death!) As might be expected, it does not therefore seek for honey secreted in the depths of flowers. Loving this food, however, as well as do its cousins, the Death's-head is frequently to be found in hives feasting on the sweet substance, as it lies stored therein. As, however, only a few individuals can be supposed to derive nourishment from this source, it remains to be discovered in what other places, and on what other substances, the insect feeds. Possibly, it may suck the juices of over-ripe or bruised fruit, as Bees are well known to do in a scarcity of honey.

---

* So called from the sphinx-like appearance of the Caterpillar when at rest.

This group contains (in the Death's-head Moth), the largest species of British Moths.

The comparatively common Humming-Bird Moth, somewhat resembling the Sphinges in general appearance, belongs also to this group, as do some remarkable *clear-winged* Moths, in which the scales, thickly planted along the margins of some of the nerves, are wanting on the membrane, which accordingly gives to these insects a curious bee-like or wasp-like appearance.

The same group contains the beautifully-coloured Burnet Moths, in which deep glossy bluish green, or greenish blue and deep bright crimson are the prevailing colours.

Most of these insects, excepting the true Sphinges, fly by day. The group contains about thirty-eight species.

The second group is Bombycina, and, with all the remaining groups, has the antennæ tapering, fine and thread-like, sometimes with a deep double feather-like fringing, sometimes only slightly fringed, sometimes simple.

Bombycina contains several stout-bodied Moths, and it is among these that the larvæ, whose silk has become an article of commerce, are found. The common Silk-worm Moths, with their beautiful feathered antennæ—the large spotted Leopard-Moth,—the pencilled grey Goat-Moth, (named from the smell of the larvæ,)—the Buff-tip (Pl. X. f. 5.)—sometimes so undistinguishable from a broken twig, as it lies among the fallen leaves on the ground—the downy, large-winged Drinker, the Tussocks, the handsome little Vapourers, and their clumsy and wingless females, the Brown Tiger, with its orange and black-spotted hind-wings, the Cream-spotted Tiger, with its black and cream-spotted fore-wings, yellow and black spotted hind-wings, the delicate Ermine, and the magnificent Peacock-eyed Emperor-Moths, are all well-known examples of this group, which contains more than 100 species. The greater number of these fly by day. One family, which flies at twilight, has obtained the name of the *Swifts*, from the character of its motions.

In the Genus Psyche in this group, not only is the female wingless, like that of the Vapourer, but she is legless, antenna-less, and lives and dies within a portable case formed and lived in by the larva. There must have been some ingenuity in the naturalist who selected this Moth, of all Moths, to bear the name of Psyche!

The antennæ in this group vary: in the principal genera the antennæ are flattened in the male, if not in both sexes. Those of the Swifts are thread-like. The most common attitude of repose is with the fore-wings laid over the hind, and deflexed, as in the Tiger-Moth, the Goat, the Leopard, &c.; but some, as the Emperor, rest with the wings extended. Others, again, assume a peculiar position, allowing the under-wings to show beyond the sides of the fore-wings.

This group is rendered interesting by the habits of some of the larvæ, which will be noticed hereafter.

Most of the night-flying Moths belong to the third group—Noctuina (whence this derives its name). These insects are of smaller size than many of the preceding, but are generally heavy-looking when in repose, their bodies being stoutish, and their fore-wings narrow—concealing the broader hind-wings, which are folded beneath them. The antennæ are generally slender and simple, the thorax is sometimes crested, and at rest the fore-wings usually cover the hind, and are deflexed.

Nearly all the Noctuina are marked on the fore-wing,

more or less distinctly, with a round or oval spot, a kidney-shaped spot, and sometimes a wedge, or club-shaped spot. Certain lines also run partly across the wing.* These markings are sometimes very faint, sometimes wanting. Most of the brown, and more or less dingy, heavy, middle-sized common Moths belong to this group, while among them are some more conspicuous. The Red-underwing, a large grey Moth with red hind-wings, decorated with broad black bands, is amongst the latter; and one genus, Plusia, glitters with gold and silver. A very pretty Moth, Gonoptera Libatrix (Pl. X. f. 6), belongs to this group. It is about an inch long when the wings are closed: of a mixed grey and brickdust colour, with minute white spots. The wings are ragged-looking, and the thorax is crested. This Moth is to be met with everywhere.

The group contains upwards of 300 species.

Geometrina is the fourth group. These Moths have broad wings, and generally slender bodies. This is not without exception; but the group is well marked by the peculiarity of the larvæ, from which it derives its name. These *Geometrina* or *Earth-measuring* Caterpillars, will be described hereafter.

The Geometrina are generally more delicate and Butterfly-like than most of the preceding groups. The large, delicate, and very beautiful sulphur-coloured Swallowtail Moth, for instance, might certainly, but for the slender and tapering antennæ, be mistaken for a Butterfly. The common little yellow Brimstone Moth (Rumia Cratægata), with its spotted wings,

---

* Similar marks are found in some families of the fifth group, Pyralidina, but the smaller bodies of the latter serve to distinguish them.

and the equally common spotted Currant Moth, are instances of this Butterfly-like character. Many species are delicately coloured and marked; a tender green, white, delicate and brighter yellows, are common among them. Exceedingly delicate pencilling also prevails among the group, and the outlines of many species are most elegant; in the attitude of repose, many of these keep the wings expanded; others, as the Currant Moth, raise them over the back in true Butterfly-fashion. The larger number, however, repose in this as in other groups, with the hind-wings concealed under the fore. The females in some families are wingless.

The number of species in Geometrina is nearly equal to that in Noctuinæ.

Pyralidina contains slender-bodied Moths, differing from those of the two preceding groups in the shape of the fore-wings, which are long and triangular. In some which lay their fore-wings horizontally on their backs in repose, this is very apparent, the outline of the insect forming a well-defined triangle, rendered the more perfect by the long, sharp *snout*, which is characteristic of many of the Pyralidina (Pl. XI., fig. 2, *Hypena proboscidalis*).

One family of these snouted Moths, known as *Grass Moths*, is easily recognised. The wings are large and limp when expanded, but when at rest are folded close round the long, slender body. In this position they are amongst the most uninteresting-looking of Moths; but it is curious to watch them on a clear sunny day, sporting by myriads in a grass field so long as the sun shines, and the moment that a cloud fleets across him, settling head downwards on the grass stalks, and with wings closely folded, so as to become in an instant almost invisible. Their inconspicuous colouring—whitish, yellowish, brownish — and the longitudinal markings in the wings of some species, greatly increase their power of concealing themselves, although under our very eye.

The writer has, on a day of swiftly alternating cloud and sunshine, watched a grass field seeming literally alive with these Moths, and with myriads of blue Butterflies, and in which, a few seconds after the obscuration of the sun, a skilled eye was required to detect the presence of either, although in a hundred places seven or eight of the lately blue—now speckled drab—Butterflies were resting on one grass-stalk, and the same stalks were thickened with the close-clinging Moths.

There are many very pretty and delicate species in this group, snoutless, of somewhat Butterfly-like aspect, and reposing with the wings spread. Some of these are beautifully marbled. The "Small Magpie" (Pl. XI., fig. 1), of which the Larva feeds on nettles, is one of the commonest of these. In others, the wings are translucent, and have the lustre and colouring of mother-of-pearl. The group contains only about one hundred and seventy species.

In Tortricina, the sixth group, the more characteristic genera are distinguished by the marking and form of the fore-wings. These are broadish, and the front margin bows out from the shoulder (see Pl. XI., fig. 3, *Xanthosetia Zygæna*); some nearly triangular Moths are, however, found among them, and, as the unicolorous hind-wings indicate, these are concealed in an attitude of repose. The colouring is very sober, and the markings are generally in patches. These insects are mostly of rather small size; the number of British species amounts to about three hundred.

Tineina is a very large group, chiefly composed of very small Moths, and which has attracted much attention on account of the variety of habit among the larvæ. The Moths are slender-bodied and very fragile, their most striking features being the extreme length of the hair or scales which fringe the wings. The form of the wings varies in different families. Some are wingless or nearly wingless in the female sex. The antennæ too, short in some families and genera, are to be found in some cases of more than six times the length of the body.

The group contains fifteen families, nearly a hundred genera, and between six and seven hundred species. The common Clothes Moths, and the exquisite little families of Leaf-Miners (see Pl. XI., fig. 4), sometimes spangled and banded with gold and silver, belong to this, the most numerous family of the tribe. Many wingless females are found among species whose larvæ live in portable cells.

The eighth family, Pterophorina, is easily recognised, containing only the ten-plume Moths (see Pl. XI. fig. 5), the white species of which, common in strawberry beds, is perhaps one of the best known as well as most beautiful of insects. Their bodies and legs are very long; their fore-wings split into two, and the hind-wings into three plumes. They fly at twilight.

Alucitina contains the twenty- (more correctly *twenty-four*) plume Moths. Only one species is known in England (see Pl. XI., fig. 6)—an inconspicuously coloured insect, which, however, standing always expanded, so as to form the most exquisite little *feather-fan*, has probably attracted the attention of most persons.

CHAPTER XXII.

LEPIDOPTERA.—LARVÆ.

IF there is little variety to be observed in the habits and manners of the perfect insects in Lepidoptera, there is much that is curious and interesting in the history of their larvæ.

Their beauty alone would call attention to many species, as in the case of the Privet-hawk Moth, with its soft green hue and purple and white decorations, and above all, its dignified and *sphinx-like* carriage.

The tufted Hop-dog, with a green coat "slashed" with black velvet and a pink-tipped tail, the larva of the Tussock, is so prized and admired in the hop counties that it is common to find it in the cottage of the very poorest and most ignorant hop-gatherers. The hump-backed, two-tailed Kitten Caterpillar, pale green, pink, and grey and white; the dark-green Caterpillar of the Emperor Moth, with its gold-spangled black bands — these and many others have but to be seen to call forth the admiration of the most determined hater of "creeping things."

But, as is often the case, it is among the less conspicuous insects that some of the most interesting habits are to be found, and there is neither time nor place in which we may not meet with one or other of the Moth larvæ or pupæ under circumstances which claim our attention.

The true Naturalist (or, to use a pleasanter and larger term, the lover of nature), while in the eyes of the world a mere idler, has, of all men, the least chance or opportunity of being idle.

The Botanist and Entomologist, for instance, will hardly pass without question a blotch in a leaf, a thread-like track in a dusty road, a hole in a tree trunk, or a patch of discoloration in a wall, unless he has traced out its history, and found a reason for its being there.

And thus it falls out that the driest hedge by a dusty road-side, the oldest paling, the newest brick-wall, presents to his mind a series of what have been called "life histories," not perhaps written out in full, but indicated; and the series of familiar signs which meets the eye of the practised Naturalist give a pleasure not unlike that which the bookworm derives from the perusal of a book-catalogue or of the book-backs in a library.

Now the student of the tribe before us has especial facilities for accounting for a spot on this leaf, a streak on that, a fragment of silk clinging to a third, and a jagged hole in a fourth. To him the spot may recall the history of a little creature sheltered in a leafy tent, constructed by itself, and carried like the house of a snail; still farther sheltered by the instinct which confined its labours below the leaf. Eating and eating, first it has destroyed the under cuticle of the leaf, then the green and tender part within, even till the upper cuticle was reached; never touching this, keeping its shelter unimpaired, until, full-grown and ready for its change, it falls to the ground, where it now lies swathed in a little shroud (its former tent), and waiting the awakening. To him the undulating streak upon the bramble leaf, no thicker than a hair in its beginning, but widening, river-like, in its onward course, tells of a visit from a glorious little sylph in gold and purple robes. The sylph is gone, but she has left behind her an almost invisible living atom—an egg, which, hatched, gives birth to an equally invisible little creature—a "caterpillar." Fully furnished with apparatus for eating and for digesting, this loses no time in setting about his labour of love; and beginning to eat the green pulp *between* the cuticles, injuring neither cuticle, he is safely housed as it might seem from every danger. And now, in the widening sinuous track, our Entomologist *sees* his invisible little glutton growing fatter and fatter, and requiring more room. He can imagine the outgrown skin burst and laid aside, and replaced by a newer and larger garment from within, and this perhaps again and again, till, like his little tent-making friend, this also drops to the ground, spins a silken shroud, and takes his long winter repose.*

What is the little shred of white silk? It may be the remains of the cunningly-devised cell of a Hunting-Spider, or it may be a silken shroud spun by some Caterpillar in which to sleep the sleep from which will come so glorious an awakening—perhaps it may be both—the emptied shroud of the Caterpillar (how emptied?), now occupied as a house, a den, a "parlour," by the Spider—and, perhaps, in a fragment of pupa case, or cast Spider-skin; the whole history may be plainly read when both the living occupants are gone. Once more. This jagged

---

* If the mine be white it is empty, if but little discoloured the miner is probably within.

hole in a leaf—who made it? A *jagged* hole—if so, it is not the work of the Leaf-Cutter Bee. She would have left her token behind in a nearly circular smooth-edged gap at the edge, and infringing on the edge of the leaf, telling of a neat carpet laid down by the little housekeeper in some well-stored home—but this hole is jagged. Is it the work of the Sawfly—of the Slug—of the Caterpillar? Our Entomologist will be able to tell us.

It has been said above that there is neither time nor place in which we may not find the traces of these creatures, if not the creatures themselves. If at one time of the year we tear a handful of moss from the trunk of a tree, out drop some little brown Chrysalids; if at another we drag a tuft of grass up by the roots, there we find silken tubes, the homes of some small Caterpillars. We find them in fungi, we find them in grain, we find them in teazle-heads, in fir-cones, in rose-buds, and in fruit; and the Hymenopterist, carefully watching the insect emerging from a Gall, discovers that he has reared in it a Moth! On the face of a lichen-covered rock we see a moving fragment, and lo! a little Caterpillar, neatly encased like a Caddis-worm in a tent of lichen, is moving and feeding, safe even from the bird's sharp eye. We open our drawers, and there, oh, sight of horror! what is that streak of white silk upon the best garment—the garment laid by, too good for common wear? We look farther, what is that dusky little roll? Is it a "great coat" on a microscopic scale? It matches our best garment ominously. It moves—a head peeps out—some little legs, and away it walks! Tell not the housekeeper!—away it walks in safety from the admiring Entomologist, if eye or lens has revealed the laborious weaving of the little garment, "his, late mine."

While, however, we look admiringly on the ingenuity of this thievish little tailor, we can but gravely contemplate his morals; for the great law, "honour among thieves," is totally disregarded by him. Two individuals which, revelling on a many-coloured woollen rug, had woven themselves most exquisitely-coloured and patterned coats, were shut up together by the writer till it should be convenient to make "specimens" of them. On opening the box it was discovered that one had eaten half of his neighbour's coat, and used up the remaining half in patching his own, with much the same effect as would be produced by the mending a kilt of one tartan with pieces of another.

Yet again. In the hollowed stems of ash, lime, &c., we find the large pith-eating larvæ of the Leopard and other Moths, in numerous small plants numerous other species; while the Goat-Moth Caterpillar does not flinch from attacking the solid trunks of timber trees, in which it forms large cavities.

Some live in the leaves of plants, carefully curled, lined with silk, and *sewed up* with silken thread. Others bind together the young leaves at the extremity of the shoots of plants, and feed luxuriously on their tender substances.

Of the Leaf-Mining species alone the variety is considerable, and the individuals are abundant. If any one doubts this, let him walk three yards along the first hedge of varied foliage which he finds. First, how many white-tracked bramble leaves will he see?—how many white tracks in one leaf (never crossing or interfering with one another)? Again, what is this large pale blotch on the leaf of a wild plum or sloe? Both surfaces seem sound; but he holds it to the light, and finds all the green substance gone from within. An elm-tree overhangs: what is this dark zigzag track? what this pretty little pink stain on the sorrel at our feet? What these puckered lines on a hundred blades of grass? Why is half this hawthorn leaf brown, and dry, and thin? Does not the irregular line of black granules between the cuticles tell of the passage of a creature feeding, digesting, rejecting? Enter the garden and look up; the drooping branches of laburnum show a hundred pale patches marked like an oyster-shell in concentric lines, and fortunate is the looker-on if the author of this *disfigurement* is present, or rather, not the author, a Caterpillar, but the beautiful creature developed from that little grub (Pl. XI., fig. 4).

Not all mined-leaves, however, have been the homes of tiny Moth-larvæ. On the leaves of buttercups, primroses, holly, honeysuckle, and many others, are found mines made by various species of two-winged Flies; and here, again, minute observation is necessary. One means of distinguishing the Lepidopterous from the Dipterous mines is afforded by the manner in which the usually black, granular, excrementitious matter is deposited, forming "a continuous track" in the mines of the Lepidoptera, while in the mines of the Diptera it is scattered irregularly. "In the blotch mine of the sloe, the work of a Lepidopterous larva,* this matter is

---

* The writer finds the rules concerning the Lepidoptera mines in a MS. note taken from the "Zoologist," and with no authority affixed.

usually in a heap in the lower end of the mine; in Rhamnus Cathartica and Clematis vitalba it appears to be fluid, not granular."

In more unexpected places we come upon the larvæ of Moths. Those of the family *Galleridæ*, in the group Pyralidinæ, inhabit in large numbers the hives of Bees, where they, protecting themselves from attack by the construction of silken galleries (hardened, it is said, with wax), feed upon the waxen combs, occasioning such mischief as sometimes to destroy the hive.

Another family, in the same group, are aquatic, some living under water in cases filled with air, while others, furnished with fish-like breathing apparatus, breathe in the water itself.

The variety of food thus shown to be used by the larvæ of Moths may probably be new to the reader, who it is likely has thought of Caterpillars as exclusively vegetarian in their habits, excepting the little Clothes-Moth grubs, with their taste for hair of all sorts, whether in the form of woollen stuffs, fur, or horsehair stuffings. But there are not only species which feed on dead animal matter of very various kinds, as the hair aforesaid, leather, grease, and butter; but there are some which, whatever their natural and usual food may be, will feed also on other living insects. This is the case with the large evil-smelling Caterpillar of the Goat Moth, and the "Satellite," and it is on this account carefully watched by the collector, as it will eat even its own species. The larvæ of the Puss Moth eat their cast skins, whilst some larvæ eat their own egg-shells on emerging from them. These may be dainty, but can hardly be very nourishing food.

With regard to variation in food, it may be mentioned

here, that though some vegetarian Caterpillars will eat almost any plants, yet in most cases individual species of larvæ are confined to single species of plants, while it is noticed that nearly-related genera of insects will be apportioned to nearly-related genera of plants. Nevertheless, there are instances of larvæ, deprived of their natural food, taking with perfect content to another kind. Thus the Silkworm, properly feeding on mulberry leaves, may be kept on lettuce; and another larva, supposed to be in its natural state an eater of fungi, has so prospered and multiplied in London wine-cellars, while feeding on the corks, as to be the cause of serious injury to the stock. In this case the change of food is argued from the impossibility that eggs or larvæ, if imported in the corks, could survive the various operations which these undergo.

And now, after this most unmethodical beginning, it is necessary to turn to details which may give the reader a clue to determining the tribe, group, or family to which belong *some* of the larvæ and pupæ with which he may meet.

The first thing to be done is to divide the larvæ and pupæ of Butterflies from those of Moths.

Mr. Stainton, in his "Manual of Butterflies and Moths," says: "The Caterpillars of Butterflies may in most instances be distinguished (*i.e.*, from those of Moths) at first sight; for, excepting the Caterpillars of the first family, all the others are of peculiar forms, either spiny or with two projecting horns at the head, one on each side, or with two short tails, or fat and short, like a Wood-Louse, or with the head much larger than the segments behind."

The Caterpillars of Butterflies have *always* sixteen legs; those of Moths vary in number.

The pupæ of Butterflies may be known from those of Moths by their more or less angular form, Moth pupæ by their rounded outline (see figs. 65, 66, p. 285; and fig. 29, p. 57).

Besides this, the pupæ of Butterflies (except those of the last family, Hesperidæ), are always naked, and fastened by silk lines to some supporting object; those of Moths are sometimes naked, sometimes enclosed in cocoons, but not suspended, naked, in the same manner as the Butterfly pupæ.* They are also frequently subterraneous.

The long, fat, soft, green Caterpillar of the Cabbage Butterfly is (boiled) but too familiar an object, and will serve as an example of the larvæ of the first family, Papilionidæ.

In the second family, Nymphalidæ, the Marbled-white and the Brown Butterflies have slender Caterpillars, distinguished by a short forked tail; that of the Purple Emperor has horns rising from the heart-shaped head, and the rest are spinous.

In the third and fourth families, Erycinidæ and Lycænidæ, the Caterpillars are short, and formed somewhat like the Wood-Louse.

In the fifth family, Hesperidæ, the Caterpillars are distinguished by the great size of the head and the small size of the segments immediately succeeding it.

The Chrysalids of Butterflies are supported in two ways, they are either suspended by the tail (fig. 65), hanging perpendicularly, or are attached by the tail to

---

* For an exception to this rule, see below, among the Geometrinæ.

LEPIDOPTERA.—LARVÆ.

some object, and then supported in an upright, inclined, or horizontal position, by a silken band or girth passed round the body (fig. 66). The Chrysalids of the

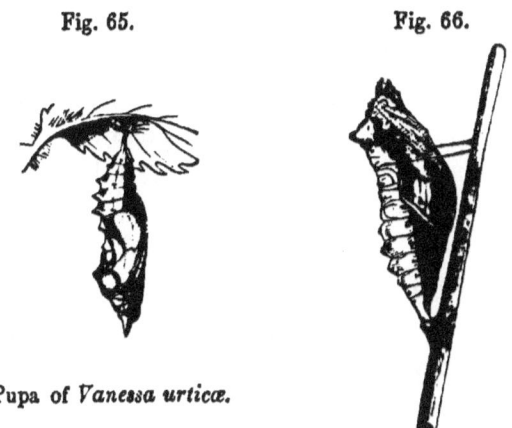

Fig. 65.

Pupa of *Vanessa urticæ*.

Fig. 66.

Pupa of *Papilio machaon*.

Nymphalidæ are suspended in the former manner, those of the other families in the latter, excepting the last family, Hesperidæ, which in this, as in many other particulars approximating to the Moths, lives during the pupæ stage in a slight cocoon, or net, as it might rather be called.

It is not possible in this small space to give rules for distinguishing the larvæ and pupæ of the different groups of Moths, as in most cases each group contains many families widely differing from each other in this particular. In a few instances, however, the reader may be enabled to decide the group or part of a group to which certain larvæ or pupæ belong.

In the first group, Sphingina, all the Caterpillars have the full number of legs (sixteen), and the greater part have one stiff and horny tail. While from the appearance of the larva of the *Sphinx* Moths is derived the name which attaches to the whole group, this is not the only remarkable-looking Caterpillar which it contains, and that of the *Elephant* Moths is even more peculiar. The front segments of this insect can be retracted or pushed forward into a tapering form like the trunk of an Elephant, and the segments immediately behind being smaller, and having large spots like eyes, gives a singular resemblance to the head of an Elephant. Some of the larvæ in this family feed on wood or pith, living within the stems of plants. The Chrysalids are naked and subterraneous.

The larvæ of the next group, Bombycinæ, present several varieties; some have sixteen legs, some fourteen, some have no visible legs at all. Some have a horny plate on their backs near the head, and some have two long tails. The Emperor Moth is garnished with bristles arranged in stars, while others are tufted with hairs, and others again, as the pretty and common Caterpillar of the Tiger Moth, are clothed with long soft fur. This last, the " Woolly Bear " of children, with whom it is almost always a favourite, has a habit of rolling itself into a ball when alarmed, and awful is the memory of nurse's legend, delivered with many warnings, of a lady round whose finger one of these rolled itself so tightly that it (*What?* Finger or Caterpillar?), that IT—had to be cut off!! In this group are some larvæ which construct cases not for their own habitation only, but for that also of the wingless female when come to maturity (see Psyche above, p. 271). This group contains the *Silkworms;* all those species whose silk is commonly used in manufacture being found here. The cocoon species of the Emperor Moth is remarkable for its elegant flask-like form.

LEPIDOPTERA.—LARVÆ. 287

In the third group, Noctuina, some have sixteen, fourteen, or twelve legs, the latter walking with a somewhat *looping* action, like those of the next family, Geometrina. In this family some of the larvæ are humped, some hairy, some have retractile heads, others are furnished with horny plates, others garnished with short and stiff hair.

Geometrina, the fourth group, is named from the peculiarity of the larvæ. These are the well-known "Loopers": Caterpillars which, having legs (true and false) only at the two extremities of their bodies, advance by nearly their whole length at each step, whence their name, "Geometrina," or earth-measurers. The Caterpillar, fixing its hind-legs to the substance on

Fig. 67.

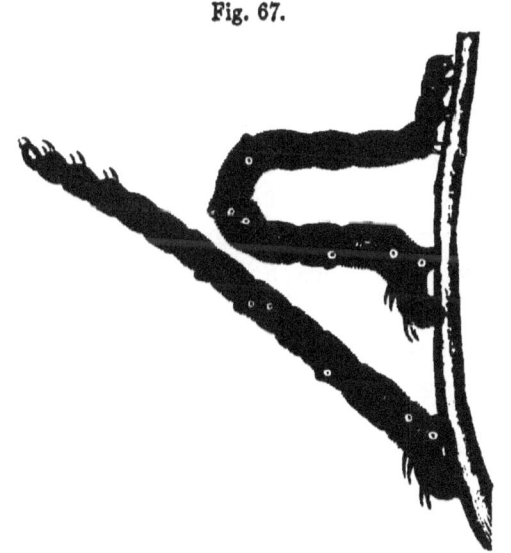

Larva of a Geometer or Looper Moth (*Ennomos*), extended and looped.

which it is walking, stretches the body to its full length, takes hold with the fore-legs, and instantly draws the hind-legs close to them: again stretching out the fore-part, the looping process is repeated, and the rapidity with which the insect progresses is very great. The application of the name is very evident, the act of *measuring* being irresistibly brought to mind by their movements.

One family of the Geometrinæ (Ephyridæ) is remarkable for suspending the Chrysalis by silken threads at the tail and round the body, like the Butterflies.

In Pyralidinæ the number of legs varies, sixteen or fourteen being the usual numbers. The larvæ are described by Mr. Stainton as having a "glassy look," and an "unusually bristly look" in the few hairs. Some of the larvæ are case-bearers, and semi-aquatic, others live on shrubs, and roll leaves wherein to change. Some live in houses, upon greasy substances, flour, &c., and others (the Galleridæ, mentioned above, p. 282) in the hives of Bees.

In Tortricinæ the legs are always sixteen, and the larvæ feed (as do many others) in leaves rolled (whence the name Tortrix), and fastened by silken threads in that position, or in stems, roots, &c.

In Tineina, the ninth and last group, we find much variety in the mode of life of the larvæ, and, as this would lead us to expect, some variety of form also. The number of legs varies from eighteen to sixteen, fourteen, and more.

The Clothes Moths, remarkable (though not standing quite alone) among Lepidopterous larvæ for their preference of animal food; the Leaf-Miners, spoken of above (pp. 277, 280), and most of the curious case-bearing larvæ, belong to this group, while in it are found some which are miners in their early state and afterwards become case-bearers, and one little species (Tinea ochraceella,) emulating the gallery-makers of the Pyralidinæ, which live in hives, constructs similar silken passages in the nests of Ants.

Many genera feed like those of other groups, naked, and on the exposed surface of leaves, or snugly sheltered within leaves which are rolled up.

The two last families, containing the Plume Moths, differ from each other in both larvæ and pupæ. The ten-plume Moths, Pterophorina, have hairy larvæ and naked pupæ, which, fastened by the tail, are, at least in one species, remarkable for activity and irritability, suddenly reversing their position if disturbed, and as suddenly returning to their former position, *head upwards*, after a few moments.

The larvæ of the twenty-plumes, Alucitina, are hairless, and the pupæ enclosed in cocoons.

TABLE OF LEPIDOPTERA.

SECTION I.—RHOPALOCERA (Butterflies).
Antennæ thickened or knobbed at the tip.
Wings in repose erect.*
Pupæ angular;* naked.*

Family 1. Papilionidæ.
Colours chiefly white and yellow.
Size from about 1½ inches to 4.
Larvæ long, naked or downy.
Pupæ secured at tail, and with a belt.
Ex.—*Cabbage and Brimstone Butterflies.*

Family 2. Nymphalidæ.
Colours generally rich and bright, or dark, or tawny.
Size about $1\frac{1}{12}$ inches to $3\frac{1}{14}$.
Legs four.
Larvæ spiny, or with two horns on head, and two short tails.
Pupa secured at tail.
Ex.—*Brown, Tortoiseshell, Admiral, Peacock, and Fritillary Butterflies.*

Family 3. Erycinidæ.
Colour brown, with tawny spots.

---

* Except in Hesperidæ.

TABLE OF LEPIDOPTERA.

Size about $1\frac{1}{12}$ inches.
Legs 4 in ♂, 6 in ♀.
Larvæ broad and short; woodlouse-like.
Pupæ secured at tail, and with a belt.

Only one British species, Nemeobius lucina.

Family 4. Lycænidæ.

Colour blue, brown, bright copper.
Size under $\frac{3}{4}$ inch to about $1\frac{1}{2}$.
Larvæ broad and short, woodlouse-like.
Pupæ secured at tail, and with a belt.

Ex.—*Blue and Copper Butterflies.*

Family 5. Hesperidæ.

Head large; antennæ wide apart.
Colour tawny or brown, generally spotted.
Size 1 inch to $1\frac{1}{8}$.
Wings in repose erect, or fore-wing erect and hind-wing horizontal.
Larvæ large heads and small necks.
Pupæ secured at tail, belted, and in a slight cocoon.

Ex.—*The Skippers.*

SECTION II.—HETEROCERA (Moths).

Antennæ various; thread-like, feathered, or spindle-shaped.
Wings in repose horizontal or deflexed; hind-wing generally covering fore-wing.
Pupæ rounded, conical; generally enclosed in a cocoon, sometimes subterranean.

* Antennæ thickest in the middle.

I. Sphingina.
Larvæ 16 legs.

Family 1. Zygœnidæ.
Colours green and brown, or green and red.
Larvæ fat, tailless.

Ex.—*Forester ; Burnet Moth.*

Family 2. Sphingidæ.
Moth large. 1¾ in. to 5 in.
Larvæ generally with horny tail.

Ex.—*Eyed Hawk Moth ; Death's-head.*

Family 3. Sesiidæ.
Wings short and broad (in *Sesia* clear); body thick.
Larvæ tailed.

Ex.—*Humming Bird Hawk Moth.*
Bee Hawk Moth.

Family 4. Egeriidæ.
Wings long and narrow, clear; body long.
Larvæ not tailed.

Ex.—*Gnat-like Trochilium.*

** Antennæ tapering from base to tip.
  a. *Body stout.*

II. Bombycina.
Fore-wings broad; body blunt at tip.

Family 1. Hepialidæ.
Ex.—*The Swifts ; Ghost Moth, &c.*

Family 2. Zeuzeridæ.
Ex.—*Wood-Leopard ; Goat-Moth.*

TABLE OF LEPIDOPTERA. 293

Family 3. Notodontidæ.
Ex.—*Puss-Moth; Prominents; Figure of* 8; *Buff-tip.*

Family 4. Liparidæ.
Ex.—*Tussock (Hop-dog Moth).*
Vapourer.

Family 5. Lithosidæ.
Ex.—*The Footman.*

Family 6. Chelonidæ.
Ex.—*Tiger and Ermine Moths.*

Family 7. Bombycidæ.
Ex.—*Oak Eggar; Lackey; Drinker.*

Family 8. Endromidæ.
Ex.—*Kentish Glory.*

Family 9. Saturnidæ.
Ex.—*Saturnia Pavonia-Minor (Emperor).*

Family 10. Platypterigidæ.
The Moths are small and slender; wings generally hooked.

Family 11. Psychidæ. Female wingless.
Larvæ carrying a case.

Family 12. Cochliopodidæ.
Larvæ Woodlouse-like; legless.

III. Noctuina.
Fore-wings rather narrow; body pointed.
(*Fore-wings generally bearing two more or less distinct spots near middle of costa—one, nearest*

base of wings, round or oval; the other kidney-shaped, and four transverse lines.

> Ex.—*Dagger; Common Wainscot; Satellite; Gonoptera Libatrix* (Pl. X. 6).

b. Body slender.

IV. Geometrina.

Fore-wings broad.
Larvæ with ten legs, walking in loops.
(*Fore-wings generally bearing a dark central spot between two dark lines.*)

> Ex.—*Oak Beauty; Large Magpie (or Currant Moth); Carpets; Pugs.*

V. Pyralidina.

Fore-wings long and triangular; much longer than hind-wings.
(*Fore-wings in some cases bearing the same marks as in Noctuina.*)

> Ex.—*Snouts* (Pl. XI. 2); *China Mark; Pearls; Small Magpie* (Pl. XI. 1).
> The Galleridæ.

VI. Tortricina.

Fore-wings rather broad; costa much curved at base.
Colour often in patches.
Larvæ mostly leaf-rollers.

> Ex.—*Green Tortrix; Zanthosetia* (Pl. XI. 3).

VII. Tineina.

Fore-wings long, with very long fringes.
Larvæ leaf-miners, case-bearers, &c.

> Ex.—*The Clothes-moths; Leaf-miners* (Pl. XI. 4), &c.

TABLE OF LEPIDOPTERA. 295

VIII. Pterophorina.
Fore-wing slit into two long feathers.
Hind-wing into three.
Ex.—*Strawberry Plume Moth.*

IX. Alucitina.
Fore-wing and hind-wing each slit into six feathers.
Ex.—*Twenty-plume Moth.*

CHAPTER XXIII.

HOMOPTERA.

THE order HOMOPTERA will best be brought before the reader by the mention of two familiar insects which it contains. These are the "Cuckoo-spit" insect, or "Frog-hopper," and the common green Rose Aphis. A very slight examination of these will show the characters of the order, and the points of difference between it and others.

First, then, to take the Cuckoo-spit (*Aphrophora spumaria*, Pl. XII., fig. 2), we see a little hopping creature, with fore-wings of a thickened texture, and placed when at rest in a shelving or roof-like position. So far it agrees with the Grasshoppers and Locusts in Orthoptera.

Next take the winged Rose Aphis. Four delicate membranous wings, united in flight by hooks, at once suggest the order Hymenoptera; but (setting aside all other characters to be presently described) look to the mouth in either of these insects, and it at once appears that, there being no biting jaws, but a sucking apparatus in the shape of a tubular rostrum or proboscis, it must belong to the second division of the order, consisting of *Sucking insects*.

Now, the only other order with which there is any excuse for confounding Homoptera, is that which follows it—Heteroptera; also an order containing insects with thickened wings, and, like Homoptera belonging to the Sucking section, like it possessing an evident beak or rostrum. And here the wings themselves, apart from all considerations of veining (which is very various in both these orders), afford sufficient distinction.

HOMOPTERA (from ὁμοῖος, *homoios = alike*, and πτερὸν, *a wing*) contains Sucking insects in which the fore-wings, whether thickened or membranous, are *of a uniform texture throughout* (see Pl. XII., figs. 1 to 6). Thus they may be thickened, as in the Frog-hopper (and so differ from the hind-wings, which are clear), or, with the hind-wings clear, and consisting throughout of thin membrane, as in the Aphis. In HETEROPTERA (from ἕτερος, *different*, and πτερὸν, *a wing*) a reference to Pl. XIII. will at once show that the fore-wing displays two distinct textures.

This, however, is not the only nor the chief difference between Homoptera and Heteroptera (although the distinctive names are derived from it), and indeed, while this was considered as the chief distinction, the two were combined in one order under the name of Hemiptera.*

The remaining characters of Homoptera are as follows; those of Heteroptera will be found in their place farther on:—

The insects are stout-bodied, sometimes with very long, but generally with short, awl-like antennæ, from

---

* The reader will do well to remember this, as the name Hemiptera frequently occurs in books both old and comparatively new, and might cause some confusion in his mind.

the last joint of which springs a bristle. The mouth is peculiarly placed, being very far back in the head, so that the proboscis springs from that part which, in a man's head, would be represented by the under-side of the chin, near the breast. The proboscis consists of the labium, which forms a jointed sheath for the slender bristle-like mandibles and maxillæ, and also a canal for the passage of the juices upon which the insect lives.

The wings usually rest in a shelving position, not overlapping one another, but to this there are exceptions.

Most species leap, but their legs are small, and do not resemble those of the Leaping Orthoptera (Grasshoppers, &c.), nor the thickened legs of the Leaping Beetles and other insects.

The pupa is active, and larvæ, pupæ, and imago much resemble each other, especially in the case of such as have wingless females.

The females have a point of resemblance with the Hymenoptera in the possession of an ovipositor, which in some species is a beautiful combination of a sawing and boring tool, holding a place not inferior to that of the Sawfly saw in the mechanism of insect anatomy.

All the insects in this order are terrestrial, and live upon the juices of plants, to which they are extremely injurious.

Many cover themselves with substances exuded from the body, and which in some cases entirely conceal the insect. Thus the Cuckoo-spit derives its name from the mass of froth so commonly found on plants in the spring, and in which the larva is enclosed. The French attribute the production of this froth to frogs (*crachat de grenouille*), and the name Frog-hopper is supposed by Mr. Westwood to have the same origin, though it may be a question whether it does not arise from the hopping, frog-like motions of the insect, which, when come to perfection, no longer inhabits the frothy nest.

Among the Aphides and their congeners, some species conceal themselves with tufts of a woolly or cottony substance, exuded, like the Cuckoo-spit froth, from their bodies; others slightly powder themselves over, or entirely bury themselves in a fine meal produced in the same manner; while others, again, as the Scale insects—relations of the foreign Cochineals—exude a secretion which will be spoken of later. Some Aphides have another mode of concealment, forming gall-like excrescences upon trees, within which they live.

Homoptera is subdivided into three sections :—

1. TRIMERA, in which the tarsi have three joints; the antennæ are very small and awl-like; two or three ocelli are generally present, and the fore-wings are sometimes thickened, sometimes clear.

This contains the Cicada, an insect related to the foreign Fire-flies and the Frog-hoppers.

2. DIMERA, in which the tarsi have two joints, and the antennæ are considerably longer than in the former sections.

This contains the Aphis family, and insects not unlike them, the Psyllidæ and Aleyrodes, tiny little white Moth-like creatures.

3. MONOMERA, in which the tarsi have only one joint, and which contains the curious Scale insects.

Cicada, in the first family in TRIMERA, consist of singular-looking insects, with wide head and thorax and a triangular abdomen. The wings are beautifully clear and distinctly veined, the antennæ are placed between the eyes, and are of six or seven joints, and there are three ocelli.

Cicada Anglica (Pl. XII., fig. 1), the only English species, is a rare insect, but there are smaller species as beautiful and more common; but while in England the Cicadæ do not form a conspicuous family, they are in some foreign countries rendered prominent both by their numbers and by the deafening noise which, when congregated together, their combined efforts are capable of producing.

The Cicada is, as has been mentioned above (Introd. p. 9), an eminently musical insect. In the Brazils it is said to "sing till it bursts," an idea arising from the number of split pupa-skins found under the trees frequented by these insects.

In America (where it is commonly called the Locust) it is less esteemed for its powers, if we may judge by the following extract:—" One of your Spa-fields meetings can give you a faint idea of their incessant and unmusical cheering and noise. If Hogarth had known these Locusts, he would have placed them about the ears of his enraged musician. Knife-grinders and ballad-singers would have been lost in their din."*

The musical instrument is neither in the wing-cases, as in the Cricket, nor in the legs, as in the Locusts; but is placed within the abdomen.

The ovipositor of the Cicada is to the full as remarkable an instrument as that of the Sawfly. It is a horny borer (*f, a*) composed of two thick blades, which may be called either saws or files (*b, c, d*) and which, running in

---

* "Journal of Science and the Arts," vol. vi. 1819.

the grooves of a supporting plate at the back (*x*), play alternately upon the wood to be bored for the reception of the Cicada's eggs. Some of this family leap.

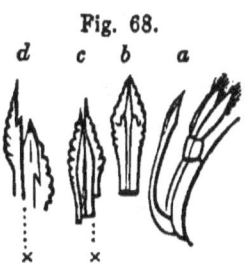

Fig. 68.

Ovipositor of Cicada. (Taken from Westwood.)
a. The borer.
b. Do. more highly magnified, seen from beneath.
c. Do. do. from above, one blade slightly protruded.
d. Do. do. blade fully protruded.
× The supporting plate at back.

In the remaining Trimerous insects the antennæ are of three joints only, and the ocelli are two in number.

The Fulgoridæ, Fire-flies or Lanthorn-flies of hot countries, find some small representatives in England (about forty or fifty species), but none of them are luminous. They may be recognised by the position of the antennæ, which are inserted below the eyes. Most of them have legs fitted for leaping. The young microscopist will find some species (as Cixius cunicularis) well worth seeking for the sake of their delicate beaded-veined fore-wings. One genus in this family (*Delphax*, containing nineteen species) might occasion some difficulty to the young student, as most of the species are usually found with only the basal half of the fore-wings developed, and wanting the hind-wings altogether. Mr. Westwood states, however, that the wings are sometimes found fully developed in hot seasons.

Next in this section come the Cercopidæ, containing

the Frog-hopping insects and others (see Pl. XII., figs. 2, 3, 4). Among these are some of very singular form, the front joint (prothorax) of the thorax being extraordinarily large, and forming sometimes a huge helmet, sometimes a large strangely-shaped shield covering the whole body. The strangest of these forms are chiefly foreign, but one species common in the New Forest is sufficiently remarkable. In this, Membracis cornuta (see Pl. XII., figs. 4 and 4 a), the prothorax forms a long, pointed, curved, and keeled process over the back, while on the shoulder it shoots out into two sharp triangular and prominent horns.

A small insect in this family sucks the juices of ferns, causing large patches of discoloration.

In the second section, DIMERA, the wings are always clear,* the antennæ sometimes of considerable length. The first family is Psyllidæ, which consists of insects very like the Aphidæ in appearance and habit, but differing from them in certain respects.

Both Psyllidæ and Aphidæ have three ocelli placed in a triangle, and long, or moderately long, slender antennæ. The wings in both are carried in a shelving roof-like position, and the range of size in the species is about the same. In both families there are species which cover themselves with a woolly secretion.

The Psyllidæ may be distinguished from the Aphidæ by their power of leaping, their very large thorax, their short rostrum, and antennæ of ten joints. The female has a visible ovipositor; the male, several small upright

---

* The wings of Aleyrodes lose the clearness of their appearance through being covered with a white mealy substance, but the membrane itself is transparent and not thickened.

appendages which garnish the upper side of the abdomen; while the Aphides are furnished in both sexes with two lateral tubercles.

The habits of the two families seem to be very similar, and indeed many of the insects commonly called Aphis are really species of Psylla. They are common on apple, pear sometimes, and birch-trees, and Psylla Buxi sets its mark on nearly every box-tree we examine, by shrinking the terminal leaves into a concave form, so giving a budlike appearance to their clusters.

Of all English Homoptera, the Aphides, or *Plant-lice*, are the most destructive and the most universally prevalent. Their attacks sometimes convert a turnip field into an offensive mass of decay, injuring, and at times destroying, whole crops of all kinds, whilst not sparing even the single little potted geranium in a garret-window. In the hop countries they form a considerable feature in the statistics of produce, and the hops would be fairly exterminated by the Aphides if it were not for several enemies already spoken of elsewhere. The Hymenopterous Ichneumons, tiny enough to be hatched, reared, and brought to perfection within the small body of an Aphis, leave the proof of their numbers in the brown swollen Aphis-skins which we may find abounding in any plant frequented by the Aphides. These displaying one small circular hole in the abdomen, tell us of the exit of the little creature which was reared and fattened in the wonderful laboratory in which vegetable juices were transformed into animal food for his sole use and benefit.

The larva of the Lacefly is another enemy, not however plentiful enough to make much havoc among the legions of the Aphis. The Syrphus larva does the gardener good service upon rose-trees, &c., being exceedingly voracious. But the great adversary of the Aphis, and one which, like itself, occasionally makes its appearance in countless swarms, is the Ladybird. This, with its larva, is so considerable a check upon the Aphis, that it is wonderful that the hop planters have not learned to add notices of the appearance of these insects to their reports on the fly, as an indication of the help to be looked for from them, either—according to the time of year—in the present or the future season.

The hop-growers, acting upon a principle all the bearings of which they probably do not fully perceive, check the production of the Aphides by a change of crop. The success of this plan is owing to the fact of certain species of Aphides feeding only on certain species of plants, so that the children of the Aphis which flourishes on the hop must starve upon the different plants which take its place. Yet, even while acting upon the experience of this fact, it is difficult to convince the rustic mind that it is a fact. The writer once came upon a gardener intent on cutting down a fine sycamore because it covered a neighbouring morella cherry-tree with blight. In vain was the plea brought forward, "the sycamore 'blight' can't live on the cherry." "But there is the cherry all covered with sycamore blight." It might have been asked, "How do you known that the sycamore is not covered with cherry blight?" That might very probably have been triumphantly answered by, "Because the cherry-fly is black, and the sycamore-fly green." But let philosophers say what they will, it is not always as long from the lion's tail to his head as it is from his head to his tail; and, pleaded on the other side, the argument had no weight.

Whilst, however, the attacks of each species of Aphis are confined to one species of plant (or, when more than one, to allied species), there are sometimes two or more species of Aphis found on one kind of plant. No part of a plant is secure from their attacks. They live not only on the exposed parts, but under the bark of trees, and upon roots buried under ground; they have even been found within the heart of apparently sound fruit.

The gait of the Aphis, except when upon the wing, is a slow creep, but some species have (in common with the Psyllidæ) another mode of locomotion which makes up to them for the want of wings. This is similar to that practised by the "Aëronautic," or Gossamer Spiders, which throw forth long silken threads, and by this means are enabled to float in the air currents to great heights and distances.* The Aphides mentioned above as extruding tufts of cottony substances, have been observed floating in the same manner, and this is probably one of the causes of the sudden appearance in fresh places of swarms of these insects. The apple-tree blight, known to us as the "American blight," and which sometimes nearly ruins the orchard, is one of these. The name has been given under the belief that it has been imported from America, but the Americans retort the accusation upon Europe, and Mr. Harris, an American writer on "Insects Injurious to Vegetation," states that it is rare in his own country, and is supposed to have

---

* The Gossamer Spiders and Aëronauts are not of any particular genus or species, the young of many kinds floating in the air by means of these threads, and any Spider which throws out long floating threads of "gossamer" being a "Gossamer" Spider.

been introduced with fruit trees from Europe. It is of recent introduction in England.*

Yet one more mode of transport has been discussed. Sir James Ross, in his appendix to " Parry's Narrative,"† says that living Aphides were found in floating ice in the Polar Sea 100 miles distant from land, and so far North as $82\frac{3}{4}°$. Resembling a species to be found on the fir, it was conjectured that "the floating trees of fir that are to be found so abundantly on the shores and to the northward of Spitzbergen, might possibly be the means by which this insect has been transported to the Northern regions." It was never seen on the wing.

When large swarms of winged Aphides have suddenly made their appearance in the air, it seems probable that (as is frequently the case in the migrations of other insects) the wind may have been mainly instrumental in conveying them.

It certainly has the credit of so doing, and the farmers commonly believe their appearance to be consequent on a north-east or east wind. So Thomson—

> " For oft engendered by the hazy north,
> Myriads on myriads, insect armies warp
> Keen in the poisoned breeze; and wasteful eat
> Through buds and bark, into the blackened core
> Their eager way."

Another substance which oozes from the tubercles at the end of the body of the Aphis is the well-known honey-dew, a favourite food of Bees, Ants, and other insects, and of which Pliny says that it is " engendered

---

* The best mode of cure is said to be covering every patch of white cotton with warm size. Whitewash is very commonly used in our orchards, and with some effect.

† See "Spitzbergen and Greenland" (Hakluyt Society), p. 165, *note*.

naturally in the air," while another philosopher gives it credit for breeding the very insects which produce it.

The presence of Aphides may be detected in cases where the insects themselves are concealed, by various effects produced upon the plants infested by them. Some species cluster on the under-side of leaves, either forming numerous little concave nests in the under-side, while the upper rises into corresponding convex excrescences, or curving the whole leaf into one mighty dome in which many hundreds live. Others, sucking the juices from the stems of plants, cause contractions and distortions of various kinds. Others, again, form large gall-like excrescences upon various parts of plants. Of these the pear-shaped sacs on the leaf-stalks of the Lombardy poplar are a common example.

A very curious little insect, supposed to be the young of some species of Aphis,[*] may be found on the underside of maple and sycamore leaves, generally (except when numerous and scattered all over the leaf) sheltering itself in the angle formed by two veins. To the naked eye it is a minute green, or brownish-green scale; under the microscope it is one of the most singular-looking creatures possible. This atom—rejoicing in the names of Chelymorpha phyllophorus, or "the leaf-bearing tortoise-shaped," and Phyllophorus testudinatus, or "the tortoise-like leaf-bearer" (and if a third variety could be formed by twisting the name any other way, the insect is quite worthy of the honour)—is a flat, tortoise-shaped green insect, bearing on its head a crown or tiara composed apparently of four beautiful *leaves*, as clear as glass, and delicately

---

[*] It much more nearly resembles the young of some other insects in the Order.

veined; two smaller leaves coquettishly decorate a prominence on the base of each antenna, slender leaflets fringe the first and second pairs of legs, and the abdomen is bordered by a series of broad leaves like those forming the tiara.

The history of Aphides is very remarkable. In the spring, numbers of fertile females, and females only, are hatched from eggs laid the autumn before. These, rapidly attaining to their full growth, but never to the possession of wings, give birth not to eggs, but to young fertile females like themselves. These repeat the same process, which occurs again and again, until at last nine generations have been produced, when, autumn having arrived, males as well as females are produced, which sometimes, but not always, develope wings; the usual pairing takes place, the female lays her eggs and dies, and from these eggs the next year's series of generations is produced.

The true pupa may be known from the permanently wingless female by its possessing the rudiments of wings. The underground species of Aphis never develope wings.

The third family of Dimera contains only the genus Aleyrodes; pretty little insects already spoken of as so covered by a fine white dust as to have the appearance of tiny Moths (Pl. XII., fig. 6, 6 *a*). The wings are carried nearly horizontally when at rest. The abdomen has none of the tubercles or other appendages common in this order. The head is remarkable as having four eyes, or, to use more scientific language, as having the eyes "parted in the middle," as are those of Gyrinus (see p. 70, fig. 32), and some few other insects. There are only two British species of Aleyrodes. They are about $\frac{1}{8}$ inch in length.

The larva is a flat, scale-like insect, and the pupa, which is quiescent, remains covered by the larva skin.

It is calculated that the descendants of one pair of Aleyrodes may amount to 200,000 in a single year, the little patriarchs possibly living to see them all!

The third and last section of Homoptera, MONOMERA, contains only the curious family known as Scale insects, Bark-Lice, or Mealy-Bugs.

The reader may often have remarked, fixed, limpet-like, on the stems and branches of vines and other trees, a convex brown scale of the size and shape of a small Ladybird, and from under the edge of which a whitish substance appears, but with no sign of head, legs, horns, or even of rings or joints. This is the dead body of a mother Coccus, or Scale insect, and on its removal from the tree the whole convex space below it will be found occupied by the white mealy exudation resembling that produced by some of the Homopterous insects, embedded in which are numerous active young Cocci with two long tails.

"By the end of July the young quit the body of their parent, and ascend to the extremity of the young branches; there they affix themselves by their rostrum, gradually increase in size, and lose their anal setæ, as well as their former activity. In this state they remain through the winter, without any diversity of appearance indicative of the sexes; and it is not till the following April that this is first perceived, by the further increased growth of the females, and by the males assuming the pupa state, which is quiescent, with the limbs arranged upon the breast, the fore-legs being directed forwards—a peculiarity not occurring in other insects."* The males

---

* *Coccus Aceris.* From Westwood's Introduction.

undergo their final transformation, and become winged about May; and by the end of June the females, which never obtain wings, but, on the contrary, become less and less perfect and insect-like as they approach maturity, are found in the shapeless state already described.

The males are comparatively perfect insects. Only one pair of nearly nerveless wings, however, is developed, which they carry overlapping each other, and the mouth, in the young and the females so powerful an instrument for mischief, is in them in a rudimentary state. The male has two long bristle-like tails resembling those of the Ephemera. They are necessarily less frequently observed than the larger and stationary females, but at the end of May or beginning of June specimens may easily be found which have not yet taken wing. These are to be found sheltered under a tiny white scale-like cocoon, from which the insect's two slender white tails project, and on raising which the insect is exposed. One of the most common English species is the Coccus of the sycamore, on the branches of which these cocoons are plentiful.

Many of the Cocci, probably imported with the plants on which they live, infest the foreign trees in our hot-houses and conservatories, as the pine-apples, orange-trees, oleanders, &c. Others, natives of England, are extremely injurious to apple and other trees, upon which they multiply to such a degree as to kill them outright. Hence, in England, the Cocci, whether native or imported, are known only by the injuries which they inflict.

Abroad the case is far different, and the Coccus or Cochineal insect is a most valuable article of merchandize. So precious is it indeed, in proportion to its bulk, that it is not unfrequently used in commerce in the place of money, changing hands several times, and making many journeys before arriving at its final destination.

Within a few years of the present time the brightest and only fast red dye was obtained from this insect, which was famous even in the times of the Greeks and Romans. Many species, from many countries, and differing greatly in value, have been used in dyeing, the most valuable of all being a Mexican species living on one of the Cacti, and which may be met with in English hothouses.*

The Lac (called shell-lac, stick-lac, &c., according to the manner of its preparation), which is used in the making of sealing-wax, different varnishes, as Japan, the lacquer used on metal, and of the pigments known by the name of "Lake," is produced by an Indian species of the Cochineal family, and is another valuable article of commerce.

The Cochineal insect is used also in medicine, both as a remedy and for the purpose of colouring other preparations.

The little seed-like, narrow brown scales, so common on the rind of oranges, are the remains of a species of Coccus.

---

* The insects are exported in various states, the best and commonest being that called "in grain"—*i.e.*, the insects whole, and with somewhat of the appearance of small grains. Thus, "scarlet grain of Poland;" in Spain "grana," or when broken "granilla." From this, the "*scarlet in grain*," (formerly, as has been said, the only fast bright-red dye) probably arose by degrees the application of the title "ingrain" or "ingrained" to other bright colours as the means of making them fast were discovered. A few years ago the only colours called "ingrain" were this fast red and a then new fast bright blue.

TABLE OF HOMOPTERA.

SECTION I.—TRIMERA.

Tarsi three-jointed.
Antennæ very small, three or six joints; the last bristle-like.
Ocelli generally present, two or three.
Wings: fore-wing sometimes uniformly thickened, hind-wing clear.

 * Antennæ six (seven ?) joints.
  Ocelli, three.
   a. Antennæ between the eyes.
    Ocelli on back of head.
    Insect not saltatorial.

Family 1.—Cicadidæ.
 Ex.—*Cicada.*

 ** Antennæ, three joints.
  Ocelli, two.
   a. Antennæ beneath the eyes.
    Ocelli beneath the eyes.
    Insect generally saltatorial.

Family 2.—Fulgoridæ.
 Ex.—*Cixius, Delphax.*

   b. Antennæ between the eyes.
    Ocelli on forehead or face.
    Insect saltatorial.

Family 3.—Cercopidæ.
 Ex.—*Cuckoo spit.*

SECTION II.—DIMERA.

Tarsi, two joints.
Antennæ longer than head; six to ten long slender joints.
Ocelli generally present, three.
Wings, hind and fore all clear.*
* Antennæ six to ten joints long, or moderate.
  Abdomen with appendages.
  Wings at rest shelving.
  a. Antennæ ten joints; in front of the eyes.
    Thorax very large.
    Abdomen ♀ with visible ovipositor.
    ♂ with several erect appendages.
    Wings at rest shelving.
    Face notched.
    Insect saltatorial.
  Family 1.—Psyllidæ.
    Ex.—" *Apple Aphis* " (falsely so called).

  b. Antennæ (six or) seven joints, third longest.
    Abdomen with two slender tubercles.
    Wings at rest nearly perpendicular; fore-wing much the largest.
    Insect crawling.
  Family 2.—Aphidæ.
    Ex.—*Rose Aphis.*

** Antennæ short, six joints.
  Eyes divided into two pairs.
  Abdomen without external appendages.
  a. Wings at rest nearly horizontal. Powdered.
  Family 3.—Aleyrodidæ.
    Ex.—*Aleyrodes.*

---

* Covered with powder in Aleyrodes.

SECTION III.—MONOMERA.

Tarsi, one joint.
Ocelli, none.
Wings clear.

* Antennæ of ♂ moderately long.
 Abdomen of ♂ with two long tails.
 Wings ♀ none, ♂ two, horizontal and overlapping.
 Eyes composite or in a group.
 ♀ scale-like.

Family 1.—Coccidæ.

Ex.—*Scale insects.*

N. B.—The above characters are chiefly derived from Westwood's Classification.

315

CHAPTER XXIV.

HETEROPTERA.

THE order Heteroptera has already been partially described with the Homoptera, to which it is very nearly allied.

It contains some well-known insects, both aquatic, as the Water Boatmen (Pl. XIII., fig. 1) and Water Scorpions (Pl. XIII., fig. 2), and terrestrial, as the beautifully coloured Plant Bugs (Pl. XIII., fig. 6), and the less attractive Bed Bug; besides some, as the slender, active, long-legged black Gerris (Pl. XIII., fig. 3), which, running and dancing on the surface of the water, can hardly be called "terestrial," though certainly not belonging to the aquatic section.

The characters of Heteroptera, and those in which it differs from Homoptera, are as follows :—

The wings are always dissimilar, the fore-wings being thick and horny at their fore-part, and membranous at the hinder part; the hind-wings clear and membranous throughout, and often of exceeding delicacy.

The proboscis springs from the fore-part of the head, instead of, as in Homoptera, from that part nearest the throat. When at rest it lies flat in both orders, pointing towards the abdomen. It is sometimes very long, but in predaceous species is generally short and strong.

The antennæ consist of from four to five joints, and

are long in the land species; in the aquatic they are shorter, of three or four joints, and are hidden from sight in furrows beneath the head.

Ocelli are sometimes present, and are two in number.

As in Homoptera the fore-part of the thorax is sometimes of unusually large proportions, so in Heteroptera is the *scutellum,* or little shield, a triangular plate extending over part of the abdomen (Pl. XIII., fig. 6, &c.), and which, in the Plant Bugs is sometimes so large as nearly to cover the wings.

The tarsi have never more than three joints; most of the land species give out a disagreeable scent.

The larvæ and pupæ are active, and resemble the perfect insect.

Heteroptera is divided into two sections.

1. HYDROCORISA, which contains the aquatic species.
2. AUROCORISA, containing the terrestrial.

Hydrocorisa contains two families, Notonectidæ and Nepidæ.

To the first belong the well-known Water Boatmen. Their shape is somewhat boat-like, and the resemblance is increased by the two long hind-legs, which, extended on either side, as the animal lies back downwards in the water, exactly represent a pair of oars, both in appearance and action.

The Notonecta Glauca (Pl. XIII., fig. 1) is a common insect, and certain to attract attention from its size, from its singular position as it floats with the under-side uppermost,* head depressed, tail (for the purpose of respiration) just touching the surface, and extended oars, and from the swiftness with which one stroke of these fringed

---

* Whence its name.

oars enables it to dart at the approach of danger, or in the hope of prey, from the spot where it has long lain motionless.

This creature has a strong, thick, curved, sharp-pointed and jointed beak, enclosing lancets, and will occasionally wound the hand which captures it. What chance the soft-bodied, plump little Tadpoles and sluggish thin-skinned larvæ of the water have against it may easily be imagined, and a single Notonecta introduced into an "aquarium" soon reduces almost any number of fat, black Tadpoles to the same number of colourless, empty, film-like skins.

Under the delicately-tinted wing-cases of the Boatmen are a pair of large, thin, milky-looking membranous wings, which the insect uses freely on occasion.

The larvæ and pupæ resemble the perfect insect, except in the possession of wings, which, however, are indicated in the pupæ.

The genus Noctonecta may be recognised by its three-jointed tarsi and overlapping fore-wings.

There are three species.

A little creature one-sixth of an inch long, and much resembling the Noctonecta, except in its greater width in proportion to its length, and comparative shortness of its hind-legs, is the *little boat*—Ploa minutissima. This is remarkable in the order as being an exception to the rule of the fore-wings overlapping. They are *united down the back in a straight line*. Like Notonecta Ploa, the fore-tarsi are three-jointed. In Sigara and Corixa they have but one joint. Corixa (in which genus are ten species) may be recognised by the little shield or scutellum being covered by the prothorax, and by an approach in the nipper-like fore-legs to the character to be noticed in the next family. In Sigara (containing one species only) the scutellum is visible.

All the Notonectidæ swim well and quickly.

The second family, Nepidæ, contains the Water Scorpions, very different in appearance from the large-headed boat-shaped Notonectidæ, and are to be known by their large prehensile fore-legs.

There are three genera, each containing only one species, and the reader will readily distinguish these.

Nepa cinerea, the common Water Scorpion (Pl. XIII., fig. 2), is a large, very flat, dingy, small-headed insect, with a tail composed of two long bristles, and a scorpion-like pair of fore-legs. It measures nearly an inch in length, or, with the tail, one inch and a-half. It is a slow and sluggish animal, living in the mud, and on the water-plants in ponds, and, notwithstanding its sluggishness, is a ravenous destroyer of other insects, especially larvæ. Probably its form and colours, resembling those of a decayed leaf, conceal it from the notice of its prey, and render swiftness in pursuit unnecessary.

The eggs of the Water Scorpion are shaped like little shuttlecocks, with feathers short and recurved.

Naucoris cimicoides is nearly oval in outline, with a much wider head than that of Nepa. The body is also rather more convex, and indeed the insect may easily be mistaken by a careless observer for a Water Beetle. It is brown and shining, and swims with some activity. The fore-legs (fig. 15, p. 38), are thicker, and even more cruel-looking (when in sight) than the nippers of the Nepa; but they are not usually held extended. The Naucoris has no tails.*

---

* For an aquatic insect somewhat resembling this, but with simple fore-legs, see *Aphelocheirus* in the next section.

Ranatra linearis resembles the Nepa in having a small head, two long tails, and extended prehensile fore-legs, but here all family likeness ends, for this most curious-looking creature is but a series of thickish *lines* (as its name imports). A long linear body with two long, thin tails, and four long, thin legs, are all we see except a pair of forceps, which would be long and thin too if they were not so crooked. The creature looks cruel and hungry, but where it stows all the prey for which it is so greedy is a problem to be solved. A less aldermanic figure can scarcely be conceived, unless in a family to be described a few pages later.

The second section, AUROCORISA, contains nine families of insects, most of which are altogether terrestrial; some skim the surface of the water, but do not swim in it; one species alone is aquatic.

The first family, Acanthiidæ, contains only two genera —the aquatic insect just mentioned, Aphelocheirus æstivalis, being the only species in one of these. It will be recognised by its aquatic habits and oval shape, which somewhat resembles that of Naucoris; while the fore-legs are quite simple instead of being pincer-like. It measures about three-eighths of an inch.

The genus Acanthia consists of insects which are terrestrial, but inhabit watery places, the seaside, riversides, &c. They are small, active, hopping creatures, of a flattish oval figure. One species is Acanthia saltatoria, a dusky insect with minute cream-coloured spots, the clear part of the fore-wings being creamy with dusky spots.

The second family, Hydrometridæ, or the *water-measurers*, may be known at once by their very slender figure, and their habit of skimming upon the surface of the water.

They are dark and lanky, with slender, angular legs.

One, Hydrometra stagnorum, an insect very common on all stagnant water, is remarkable for extreme slenderness, being scarcely thicker than a fine thread, and about three-eighths of an inch in length. The head is very long, the eyes are excessively prominent, and the wing-cases do not overlap. Gerris lacustris, and Gerris paludum (Pl. XIII., fig. 3), two other common species, are larger and somewhat bulkier, and are very conspicuous as they *skate* on the water with the utmost swiftness and ease of motion. Velia rivulorum is a rather less common insect, shorter and not so slender as the last, and of livelier colouring, black, red, and white

There are five genera in this family.

The third family, Reduviidæ, contains genera and species varying much in figure and general appearance, but to be recognised by the short, thick, curved beak, the neck-like form of the back of the head, and the long antennæ with the last joint very slender. The larvæ of one species in this family, Reduvius personatus, has the curious habit of thickly covering itself with dust, so as almost to lose the appearance of an insect. A similar habit has been already described in the larvæ of the Tortoise Beetle. Reduvius personatus is found indoors, and is said to prey on the Bed Bug. Some of the family do not, except under peculiar circumstances, fully develope their wings and wing-cases.

There are six genera in this family.

Cimicidæ, the fourth family, enjoys the distinction of containing the Bed Bug. There is one genus, in which are four species. These are all very flat, roundish in outline, and even in the perfect state have but the indication of one pair of undeveloped wings, consisting of two little scale-like appendages.

The Bed Bug is generally supposed to have been imported from abroad—some persons think in pine-wood from America. Southall (in Westwood) says that its first appearance was after the fire of London in 1666; but it is mentioned as having been seen in 1503. It is believed to feed upon the sap of the pine, and certainly " harbours" not only in new but also in old wood—a fact which has helped to make common the substitution of iron for wooden bedsteads. The countless hosts in which these insects occasionally appear, not by degrees but suddenly, are very remarkable, and the superstition attaches to them that this sudden appearance is portentous of a death in the house. It is very certain that they often live upon food other than the juices of animals, from an account published some years ago of a long closed and neglected building, which on being opened, was found to contain these insects in millions; they were taken out in shovelfuls, and it required the labour of days to effect anything like a clearance from them. Pigeon-houses are liable to be infected by them.

Tingidæ, the fifth family, contains nine genera of broad, flat insects, varying in many respects, but to be recognised by the antennæ, of which the last joint is as thick as, or thicker than, the preceding, and by the short three-jointed beak, which lies in a groove under the head. The species are all small.

In some genera, the shell of the thorax and the elytra are much wider than the body to be covered, and the pro-thorax goes down into a point like the conspicuous scutellum of some other heteropterous insects. Most of the family display a beautiful network on the surface of the thorax and elytra.

In one genus, Aradus, on the contrary, there is a

scutellum, and the thorax and wing-cases (which are strongly veined), are reduced to the usual size—the latter, indeed, being rather below this, as they do not entirely conceal the abdomen.

In the genus Aneurus (containing one species, *lævis*) the wing-cases are entirely without nerves. They are sluggish insects, and live on the juices of plants.

In the genus Acalypta (containing only one species, *carinatus*), the rule of Heteroptera is broken by the wing-cases meeting in a straight line like those of a Beetle; while in the genus Agramma (containing only one species, *læta*), another exception to the rule is found in the wing-cases being of uniform texture throughout.

Capsidæ contains both more genera and more species than any other family, the species numbering more than eighty. They are small, soft, delicate-looking insects, with long antennæ, which generally, but not always, are slender towards the tip, and much thicker below. This peculiarity is conspicuous in Capsus spissicornis (Pl. XIII., fig. 4), a common little dark insect, which, small as it is (about three-eighths of an inch in length), is likely to attract observation from this circumstance—the antennæ are about three-fourths the length of the body, and thick enough greatly to increase the apparent size and importance of their small owner.

Another common species is Capsus flavomaculatus (Fabr.), which measures rather more than a quarter of an inch, and may be known by its black thorax and hemelytra, the latter banded with yellow; the legs are yellow, and the membranous part of the fore-wings is beautifully iridescent. The family generally is marked by the fore-wings and hemelytra, the horny part of which is "terminated by a large triangular piece, like a stigma, the apical membrane having only one or two strong veins, curved and parallel with the tip of the wing, forming a basal semicircular cell."

Pretty as they are, the little Capsidæ, like others of their order, are first cousins to the abhorred Bug, and the family likeness makes its appearance in the disagreeable scent belonging to them, and which they leave behind on fruit, upon the juice of which they have been feeding.

The Lygæidæ are generally small and somewhat slender insects, often prettily banded and spotted with black, red, and white. Lygæus equestris (Pl. XIII., fig. 5), is one of the larger and more conspicuous species; the thorax and horny part of the wings are red banded with black, and the black membrane of the wings is beautifully spotted with white. Some are remarkable for the form of the thighs of the fore-legs, which are much thickened and curiously toothed. Gastrodes abietis, an insect about one-third of an inch long, is an example of this. The head and fore-part of the thorax, the scutellum, and half the antennæ are black; the basal joints of the antennæ, edge and base of the thorax, and legs are yellow; the wings are yellowish, spotted with red. In this family is one genus, Astemma (three species), in which the elytra neither overlap, nor are half horny, half membranous, but are thickened throughout.

Lygæidæ and the following family, Coreidæ, may be distinguished by the longitudinal veins in the membranous part of the fore-wings. In Lygæidæ these are seldom more than five in number, while they are numerous in Coreidæ. In Coreidæ too, the last joint of the antennæ is thickened, but not in Lygæidæ.

The Coreidæ, like the Lygæidæ, contain many

prettily-coloured species, and are often remarkable for peculiarities both of form and texture.

In some genera, the thorax is dilated at the sides, the wing-cases rarely entirely cover the abdomen, and the legs are long.

One genus, Rhopalus, is remarkable for the texture of the wings, the fore-part of which is nearly clear, and is outlined and crossed by exceedingly strong, thick nerves, forming cells. The lower part is numerously horned. This genus contains only one species, R. capitatus, a yellow insect about three-eighths of an inch in length, with thick, hairy antennæ.

Another species, Cymus resedæ, about one-fifth of an inch in length, with red head and scutellum, and the rest of the body yellow, is found on Mignonette.

These insects are all active, both in running and flying, and are supposed to live on the juices of plants.

The family Scutelleridæ derives its name from the great size of the scutellum or triangular shield, which overlies the abdomen. The proboscis and antennæ are long, the latter frequently consisting of five joints.

There are twelve genera, and about forty species, of which fifteen are in the Genus Pentatoma. Pentatoma rufipes (Pl. XIII., fig. 6) is a common and handsome species, but does not display the family character of an enlarged scutellum in anything like the degree in which it may be seen in other genera. Thus, in Podops inunctus, a broad, brown, beetle-like insect, about a quarter of an inch in length, and with two curious horns or epaulettes projecting from the shoulders, the large rounded scutellum nearly covers the wings.

In Eurydema oleracea, a beautiful oval insect of about the same length, the scutellum is nearly as long, but.

narrower and more pointed. This insect is of a deep blue colour, the head, thorax, and wing-cases are outlined with red, a band of red runs down the middle of the thorax, the scutellum has two lateral red spots and is tipped with red, and a large red spot occurs at the end of the horny part of each wing-case.

Cydnus bicolor is another pretty species, roundish and flat, about a quarter of an inch in length, and piebald—black and white; the wings leave the end of the abdomen exposed, showing a pattern of black and yellow.

The insects in this family are very various in colouring, many being exceedingly beautiful, and they are chiefly of large or middle size. They live principally on the juice of plants, but will also prey on other living insects.

Mr. Westwood quotes a statement that six or eight specimens of Pentatoma bidens (a quietly-coloured, yellowish-brown species, with a sharp tooth projecting on each side of its thorax), shut up in a room swarming with the Bed Bug for several weeks, completely extirpated them.

De Geer observed some females in this family accompanying and guarding their young brood as a hen her chickens, or an earwig her little earwigs.

TABLE OF HETEROPTERA.

SECTION I.—HYDROCORISA.

Antennæ short and concealed.*
Legs fitted for swimming.*
Ocelli wanting.

    A. Form boat-like.
        Hind-legs long and oar-like.
        Head about as wide as thorax.

        Family 1.—*Notonectidæ.*

        Ex.—*Water Boatmen.*

    B. Form flat and broad; or linear.
        Fore-leg prehensile.
        Head much narrower than thorax.

        Family 2. Nepidæ.

        Ex.—*Water Scorpion.*

SECTION II.—AUROCORISA.

Antennæ long.
Legs fit for running.
Ocelli often present.

---

* Aphelocheirus, an aquatic genus in the family Acanthidæ, has these characters of Hydrocorisa.

TABLE OF HETEROPTERA. 327.

A. Terminal joint or joints, or joints of Antennæ not more slender than those preceding.
Tarsi usually three-jointed.

* Rostrum long, second joint elongated and straight.
Figure oval, flat.

Family 1. Acanthidæ.

** Rostrum moderately long, four-jointed; third joint much longer than the rest.
Figure long, very slender.
Tarsi two- or three-jointed.

Family 2. Hydrometridæ.

*** Rostrum short, three-jointed, buried in a groove under the head.
Tarsi two- or three-jointed.
Figure flat, broad.

Family 3. Tingidæ.

**** Rostrum moderately long, joints nearly equal.
Figure generally narrow.
Ocelli sometimes wanting.

Family 4. Lygæidæ.

***** Rostrum long, four-jointed.
Antennæ long, often five-jointed.
Scutellum large.

Family 5. Scutelleridæ.

****** Rostrum moderately long; third joint shorter than fourth.
Antennæ terminal joint large.
Tarsi three-jointed.

Family 6. Coreidæ.

B. Terminal joints of antennæ slender.
* Rostrum short, thick, naked, curved.
  Head narrowed into a neck.
  Tarsi three-jointed.

    Family 7. Reduviidæ.

** Rostrum moderately long, three-jointed.
  Abdomen flat, nearly round.
  Wings wanting.

    Family 8. Cimicidæ. (Bed Bug, &c.)

*** Rostrum long, four-jointed.
  Body convex, soft.
  Ocelli wanting.

    Family 9. Capsidæ.

CHAPTER XXV.

ORDER XII.—APHANIPTERA.

THE order APHANIPTERA contains the family of Fleas only—insects which, as that name imports, are entirely destitute of wings.

There may seem little that is remarkable in this circumstance; apterous species, and apterous individuals of winged species, being found in all or nearly all other orders. But of the Fleas, which are considered to form an order by themselves, not a single species, British or foreign, is known to develope wings. It is true that four little scales supposed to represent these are found upon the thoracic segments, and Naturalists have observed " something like elytra," and " vestiges of wings," but anything which could be called *wings* has never been found.

Indeed, it seems that the process of development in the Flea is arrested before it comes to the wings, for it is unlike nearly all other insects (except such as the imperfectly developed female of the Glowworm) in having no distinct thorax. The body, from the head to the tail, is composed of a series of rings or plates, not soldered together in separate masses as those which form the thorax and abdomen in other cases, and the insect thus assumes rather the appearance of such a larva as occurs in the families with imperfect transformations, than that of a perfect insect.

The transformations, however, of the Flea, are not of the kind called "imperfect;" the larva is a long, footless, worm-like little grub, and the pupa is quiescent, resembling that in Coleoptera and Hymenoptera.

The mouth is formed for sucking, and is composed of the usual parts—mandibles, maxillæ, labial and maxillary palpi, and tongue, but is deficient in the upper lip. The mandibles are transformed into serrated lancets, to which the labial palpi form a sheath, the maxillæ are small, and the jointed maxillary palpi standing out in front of the head might easily be mistaken for antennæ. The real antennæ are small, curiously formed, and generally concealed. The body is compressed, the legs long and very powerful, especially in the action of leaping.

About twenty British species are known, but it is probable that many remain to be discovered, parasitic on quadrupeds and birds. Man himself, cats, dogs, bats, moles, pigeons, &c., are infested by them—generally, each by a species peculiar to itself. Each species prefers the animal to which it belongs, and it is therefore seldom or never that a Flea found upon our persons—albeit it was hatched in the hen-house, or in the rug on which our dog was lying—is either a dog's Flea or a chicken's Flea. Yet it must be confessed that there is a natural reluctance on our part to lay claim to any especial property in this insect. We can enter into the feelings of a certain Princess, when an "Industrious Flea," having escaped while being exhibited, and, as it was supposed, "taken refuge with her Royal Highness," the culprit was sought for, captured, presented to the exhibitor, and declared to be—not an educated, but a "wild" Flea—not his, but hers.

Fleas thrive especially in dirty and untidy houses, and

other places where dust, flue, particles of animal matter, &c., afford a harbour to the eggs and both food and lodgings to the larvæ. They have been found swarming at the mouths of the deserted holes of Sand Martins; and a traveller (Sir Howard Douglas?) speaks of them as so numerous in one place that if they had but been unanimous they might have pushed him out of bed.

The Chigoe or Jigger of the West Indies and South America is a species of Flea, and is far more objectionable than any of our European species, from its habit of burying itself in the skin, causing inflammation and sores which are sometimes even fatal.

A curious legend, preserved among a sect of Kurds who dwelt at the foot of Mount Sindshar, is quoted by the author of "Episodes of Insect Life," whence we will borrow it *verbatim* :—

"When Noah's ark sprang a leak by striking against a rock in the vicinity of Mount Sindshar, and Noah despaired altogether of safety, the Serpent promised to help him out of his mishap, if he would engage to feed him upon human flesh after the deluge had subsided. Noah pledged himself to do so, and the Serpent, coiling himself up, drove his body into the fracture and stopped the leak. When the pluvious element was appeased, and all were making their way out of the ark, the Serpent insisted upon the fulfilment of the pledge he had received; but Noah, by Gabriel's advice, committed the Serpent to the flames, and, scattering its ashes in the air, there arose out of them Flies, Fleas, Lice, Bugs, and all such sorts of vermin as prey upon human blood, and in this manner was Noah's pledge redeemed."

Deus Myiagrus.

CHAPTER XXVI.

ORDER XIII.—DIPTERA.

WE now come to the order of Flies, a tribe regarded with much disfavour by those who, not looking beyond the apparent evils which they occasion, are ignorant of their great importance in the economy of nature.

Not only in the ancient times already spoken of—times when a god* was summoned to disperse, as a cause of evil, creatures which were in truth its antidote; not then only, but even to the present day these most useful little creatures are thought of, spoken of, and treated as an unmixed nuisance. In Greece and Rome two thousand years ago they were looked upon as bringing pestilence; in England to-day, we hear of the "Cholera Fly," not as a Fly *coming with*, but as one

---

* See Introduction, p. 2, and the figure at the head of this chapter.

"*bringing*" disease. In some years the large swarms of so-called "Cholera Flies" have been flights of Aphides. In this case it is less easy to trace either the final cause of the presence of these insects or their connexion with the appearance of cholera (if such connexion exists), than in that of the true Flies, numbers of which, bred in and feeding on the substances which induce disease, are both dependent on the presence of those substances, and corrective of their noxious influences. We may, however, believe that certain conditions of the air favourable to the development of disease, may be equally favourable to the development of vegetable life, and consequently to that of vegetable-feeding insects.

Enough has already been said of the usefulness of Scavenger insects, and of their almost universal presence where their labours are required, to suggest the thought that where pestilence is rife, whether from careless uncleanliness or from such noxious atmospheric influences as produce the same effects, there will be found the myriads of Flies, whose office it is, in their earlier stages, to consume the deadly substances which fill the air with poison. It is easy to conceive that the heathen feasts and sacrifices, unguarded by the regulations which surrounded those of the Jews, might well call for the presence of these little guests; while in their bestowing the names of Μυῖαι, *muscæ*, on the uninvited and unwelcome human parasites, who thrust themselves upon their feasts, we may read the feelings which they, ignorant and ungrateful, entertained towards their little benefactors.

But it may be said of the carrion-eating, dung-eating, Scavenger Flies, that their purpose is apparent, and for the sake of their general utility, we are willing to condone the offences of even those who, under a mistaken view of duty, visit our larders to assist us in the removal of beef and mutton. What, however, is to be said in defence of those hordes whose aggressions touch our persons—of the blood-sucking little demons which, a small misery in an English summer, make whole tracts of country uninhabitable in tropical regions? What of the swarms of Gnats, of Midges, of Mosquitoes, of Sandflies, which might almost have prompted the thought of our forefathers—that the fallen spirits, shut out from the upper heaven because of its delights, from the lower earth because they could not there torment the race of men, are confined midway in the darkened air, where, as Flies, they surround us—numberless, filling the air—so that it is full of devils and evil spirits, "as the sonne bemes ben full of small motes, which is small dust or poudre." Is there any defence for these creatures?

Of the bloodsuckers in the race of Gnats, Midges, Mosquitoes, &c., nearly all the larvæ are aquatic, and consumers of decaying matter—Scavengers. If this be so, is not the purpose of the production of the myriads of such creatures in the unhealthy swamps of unreclaimed tropical lands sufficiently evident; and would it be straining one's view of final causes to trace the continuance of their beneficial influence on man, when, changed from a Scavenger Maggot to a bloodthirsty little Midge, our tormentor drives us from the regions as yet unsuited for our habitation?

That Flies are officious, busy, curious, there is no denying. A well-known Artist of the present day tells an amusing story of the interference of a little Housefly, in which we might almost imagine the spirit of fun and mischief at work. The painter having left the room in which he was giving a lesson in miniature painting, returned to find the carefully finished eyes as carefully unfinished again, denuded of all colour in the iris. He cast an unfriendly eye on the pupil's little brother (who had been left alone with the picture), and restored the eyes. Called away again, the *unpainting* process was partially repeated, and this time the little brother was openly accused, denied the charge indignantly, was believed by his mother, and disbelieved by the angry master. No sooner was the discussion dropped, than casting his eyes upon the miniature, the Artist observed a Housefly busily at work, delicately sucking up with his tiny proboscis all that remained of the colour employed in painting the irides.

And now, having talked for three pages about Flies, it seems time to inquire what is meant by a "*Fly*." This depends very much upon who is speaking. Being a "popular name" the "people" have a right to mean what they choose by it, and they avail themselves of this right—some meaning by it one thing, some another, some every flying insect for which they know no other name. Thus, "the Fly" of the farmer is usually the little hopping Turnip Beetle; the "Fly" of the hop-grower is an Aphis; the "Fly" of the herdsman, a Gad; while to the citizen, almost anything to be seen with wings (except pigeons and sparrows), is a *Fly*.

There are some again to whom Flies are *Flies*—one Fly —*the* Fly—the common well-known little black *Housefly*. Here at last is something definite. No, not even now; for these will at least claim their young Housefly, and their full-grown Housefly, and expect you to believe that late in the year their Housefly takes to biting you: little dreaming that the little Fly, and the big Fly, and the Fly which bites you, not only are different species, but even belong to different genera: that the little Fly never grows big, that the big Fly never was little, and that their Housefly could not bite you if he would.

What, then, are we to understand by the name Fly? It is clear that the popular sense has no sense at all— or too many senses—and yet the word cannot be spared from our vocabulary. In any Latin dictionary we shall find Musca (Fly), and the Entomologist pounces upon it, and says, it shall mean the *tribe of two-winged insects.*

Linnæus so used it, and his genus Musca, now broken up into many new genera, represented the greater number of those insects which the Entomologist now claims as Flies.

The order DIPTERA, then, is marked by the absence of hind-wings, the place of which is occupied by two small, short, hair-like appendages, ending in a knob, and termed halters or poisers. The wings are membranous, not closely veined, and are never folded. At their base, a little wing-like membrane, called alulet, or winglet, is most frequently found.

The mouth is fitted for sucking only, and its principal parts are a sucker, or fleshy tongue, familiar to us as the "proboscis," or "trunk" of the Housefly, and several fine lancet-like organs. It is these latter which, in the blood-sucking Flies, or Gnats, Horseflies, &c., are used to pierce the skin, while the fleshy tongue or sucker makes a vacuum, and draws away the blood. In fact, when a Gnat "bites" us, the truth is, that the little creature puts us through the exact process of *cupping*. The fleshy sucker is the labium, lip, or tongue, as it is variously called in this and other insects. The lancets are the upper lip, mandible, maxillæ, and palpi—or some of these—completely changed from their form as that is seen in Biting insects. These lancets, always delicate, are nevertheless comparatively strong in the Blood-Suckers; while in those Flies which live on fluids not enclosed in thick-skinned vessels, they are feeble and flaccid.

The antennæ vary greatly. Under one form there are two or three short joints, of which the terminal is large and sometimes nearly globose, with a bristle springing from its upper side or from its apex. Under another, the several joints form a more or less spindle-shaped antenna, with or without a terminal bristle, while in the Gnats, Daddy Longlegs, &c., the antennæ are long, slender, many-jointed, and beautifully decorated with whorls of hair. There are also many intermediate forms.

The tarsi are five-jointed. The commonest form of foot consists of a pair of curved claws above a pair of flat, sucker-like, hairy pads. The claws vary in form, and the pads both in form and number—there being two, three, or, rarely, none.

The larvæ of Flies are generally legless maggots of simple form—the "Gentle" used by anglers being a well-known example; but some of the aquatic species are more complicated externally, and are furnished with ornamental appendages belonging to the breathing apparatus.

The pupæ are inactive, a curious exception being found in the Gnat family, of which the pupa (aquatic) is very active, although unable to feed. The pupæ are sometimes naked, and sometimes remain enclosed in the larva skin, which either retains much of its original form or contracts into a smooth, egg-like case or cocoon. The pupa itself, whether naked or enclosed in the larva skin, resembles that of the Beetles in having the limbs separately cased.

The manner in which Flies are produced varies. Most are, like other insects, produced in the egg state; others, among the Carrion Flies, are born, not in the egg, but already grown to larvæ; while, in the case of certain Parasitic Flies, they even attain the pupa stage before exclusion.

As in all the four-winged orders of insects, some species or sexes are found wanting one or both pairs of wings,* so in the order Diptera, characterized by the invariable absence of hind-wings, the fore as well as the hind pair are sometimes wanting, and also the halteres or representatives of the hind-wings.

If it be asked how, when, as in Diptera, or exceptionally in any of the orders above named, there is but one pair of wings, they can be pronounced to be hind-wings or fore-wings?—the answer is, much in the same way as if a monstrous horse were born with only two legs, it would be decided whether these were the hind or the fore legs; their relation to other parts would settle the question.

The order Diptera is divided by marked characters into two sections, PROBOSCIDEA and EPROBOSCIDEA, but of these the first contains the bulk of the order, while the second contains only a few known species, which are all parasitic, living *in the perfect state* on the surface of the bodies of quadrupeds and birds.

The characters of PROBOSCIDEA are—*proboscis* fleshy

---

* When only one is absent, it is nearly always the hind pair.

and bilobed at the tip; *legs* of the opposite sides inserted down the middle of the thorax; *head and thorax* distinct, being connected by a neck; *antennæ* placed between the eyes.

In EPROBOSCIDEA, the *proboscis* is tubular, the *legs* are set wide apart, those of opposite sides being separated by a wide breastplate. The *head* is either sunk in the thorax, or thrown so completely backward as to be actually reversed, and the *antennæ* are partially buried in the head.

The first section, PROBOSCIDEA, is divided into two large groups, named, from the characters of the antennæ,

1. *NEMOCERA.*
2. *BRACHYCERA.*

To this is added a third, *HYPOCERA*, which consists but of one genus, containing only a few small species.

To *NEMOCERA* belong the Gnats, Daddy Longlegs, and others, having long and slender thread-like antennæ of several joints, numbering from six to sixteen, and fre-

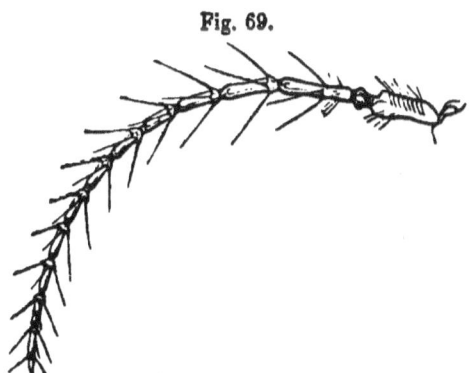

Fig. 69.

Antenna of Tipula.

quently very beautifully decorated with whorls of long or short slender hairs (see fig. 59, and Pl. XIV., fig. 1). The Flies of this division are nearly all to be recognised with ease by their slender form, small head, high thorax, and long and delicate legs, which are extended downwards and backwards during flight.

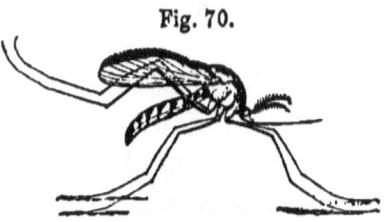

Fig. 70.

Female Gnat (*Culex Pipiens*) at rest.

In *BRACHYCERA* the antennæ are comparatively short, very often consisting of only three joints of unequal size, from the last of which a bristle or delicate feather usually springs (see Pl. XIV., 6, *a*; XV., 4, *a*; XVI., 3, *a*, 5, *a*). The antennæ sometimes are longer and have more joints (see Pl. XIV., 2, *a*, 4, *a*)—sometimes as many as ten—but these, after the third, are usually more or less consolidated into one, and have a character different from that of the distinctly articulated antennæ in Nemocera (see fig. above). By far the greater number of Flies belong to Brachycera.

In the small section *HYPOCERA*, the antennæ much resemble those of the Brachycera, but are differently placed, being low down and close to the mouth. The character of the mouth, in which the lancets are not developed, with some other characters to be named in their place, also help to distinguish it from Brachycera.

The wings in Brachycera are usually characterized by the posterior nerves forming several perfect cells.* In most families the membrane of the wing forms a larger or smaller lobe in the axil, which lobe is very small

---

* See figures in table of Diptera.

or wanting in Nemocera; and the winglets, undeveloped in Nemocera, are often conspicuous in Brachycera.

The habits of Flies both in the larval and perfect states vary much. Of the larvæ, many are purely aquatic, as the well-known active little Gnat larva common in all pools, ponds, and tanks; or live in wet mud and filth, as the useful "rat-tailed" larva of the Dronefly. Others live in the earth, feeding on decayed matter, or on the roots of plants; while some, as the "Gentle" of the angler, are deposited, already hatched from the egg, in the carrion which it is their office to consume.* They are found feeding in almost every part of almost every kind of plant, and a large number live in unhealthy growths upon plants—similar to the galls of the Hymenoptera— occasioned by their own presence. Some few feed upon other living insects. Some, again, are parasitic, inhabiting the nests and feeding on the food of other insects; while others, as the Gadflies, are parasitic within the bodies of quadrupeds.

The variety of food chosen by various Flies when arrived at perfection is nearly as great as that of the larvæ. While perhaps the greater number of species feed on the honey in flowers, and on the various juices of plants, others, as the Gnats, Horseflies, &c., suck the blood of men and animals whenever an opportunity offers. There are also true parasites among the Flies— species which, winged or wingless, live constantly, in the complete state, upon the bodies of birds and quadrupeds. Besides these, many Flies are predaceous, attacking other insects and draining them of their blood, and some few live upon the fetid juices of carrion.

---

* The value of this provision in the saving of time, when the object is to check the evil effects of putrefaction, needs no remark.

As, however, is commonly the case in insects, predaceous or vegetarian habits in the one state do not imply a continuation of the same habit in another, and we not seldom find the predaceous larva followed by a vegetarian Fly, and the reverse. There are, however, a few rules which appear to prevail, and which are worthy of note.

In the first division of Proboscidea, Nemocera, it may generally be observed that the Blood-sucking Flies (*i.e.*, Gnats, Midges, &c.) are those which proceed from aquatic larvæ, the terrestrial larvæ which live chiefly on fungi and other plants, living or decayed, producing flower-loving Flies.

In the second division, Brachycera, on the contrary, the Flies proceeding from aquatic larvæ feed on the honey in flowers, &c., while the blood-sucking Horsefly, the predaceous Empis, and carrion-eating Flies are terrestrial in the larva state. It is in this division that the parasitic *larvæ* are found, of both kinds of parasitism, in the nests and bodies of other insects, and in the bodies of animals, and these larvæ produce Flower-loving Flies.

The rules just given are not to be considered as without exception, even among insects whose habits are known ; and it must be remembered that there is so large a number of insects of whose habits in one or both states we are as yet entirely ignorant, that it is impossible to lay down general rules in those matters which may not hereafter prove to be valueless.

The Parasitic Flies—Flies, that is, which are parasitic in their perfect state upon quadrupeds or birds—are found only in the second section, Eproboscidea, and, with the peculiarities of their transformations, have been already described.

The number of the Diptera is so large, that to describe

the families only in a manner which may enable the young student to refer species to them, is all that can be attempted here; and the reader will find it necessary for this to make himself acquainted with the figure of a wing and its principal veins and cells given at the end of the table of Diptera.

The *NEMOCERA* have been divided into two families, *Culicidæ*, in which the proboscis is very long, and *Tipulidæ*, in which it is short—the Gnat (*Culex*) being the type of the former, the Daddy Longlegs (*Tipula*) of the latter. This marks off the Culex family with distinctness, but the shortness or length of the proboscis and membership of this family, are no certain indications as to the blood-sucking habits of the species. The group is now divided by Mr. Walker into ten families* (Culicidæ and Tipulidæ being two of them); but even so the Blood-suckers are not entirely separated from the Vegetarians, one genus in Culicidæ itself not being blood-sucking, while Blood-suckers and Vegetarians are mixed in other families.

Indeed, this habit is not so important a difference as we (from an interested point of view) might be inclined to consider it; for it is evident that of the myriads of Gnats and Midges which are produced in uninhabited swamps and forests, the greater part must die without having ever tasted blood, but having fed, if indeed food is necessary during their short lives, on the juices of plants, which their piercing and sucking probosces are as well suited to obtain as the fluids contained in the veins of animals. The pupæ in Nemocera are naked.

---

* Mr. Walker, indeed, reckons eleven families in Diptera, the Fleas forming one of these; but in the present work these form an order by themselves.

Mr. Walker divides *NEMOCERA* into the following families:—
1. Mycetophilidæ.
2. Cecidomyzidæ (Gall-gnats).
3. Bibionidæ.
4. Simulidæ (Sandflies).
5. Chironomidæ (Midges).
6. Culicidæ (Gnats).
7. Phlebotomidæ.
8. Heteroclitæ.
9. Tipulidæ (Daddy Longlegs, Craneflies).
10. Rhyphidæ.

In the three first of these familiies the larvæ are terrestrial, living on fungi, roots of grain, &c., and dead organic matter. The perfect Flies are not blood-suckers.

1. Mycetophilidæ. These, as their name denotes, live chiefly upon fungi, partly also (as is common with fungus-eating insects) on decaying vegetable matter of other kinds. They are little, active, *hopping* creatures, in general appearance like minute and beautiful Gnats, from which, however, the shortness of the proboscis and the comparative shortness of the legs serve to distinguish them. They are further to be distinguished as usually possessing ocelli, two or three in number—Bibionidæ, and, rarely, Cecidomyzidæ being the only other families of Nemocera with ocelli.

The wings are without the discal areolet.

2. *Cecidomyzidæ.*—These are also very small and exquisitely beautiful Gnat-like little insects, with glittering rainbow-coloured wings, and often with long, slender, and decorated antennæ. They seldom have ocelli. The Gall-making larvæ are found in this family, while others feed in the seed-vessels, flowers, leaves, &c., of living

plants, and some on decaying wood and even in the woody galls formed by other insects.

As was observed above concerning the Lepidoptera, so here it may be repeated of Diptera—that a study of the order, and especially of this family, will inform the young naturalist of the history of many of the excrescences, monstrosities, and, in some cases, decorations, which he cannot fail to observe in any country walk. The larvæ of some of the Cecidomyzidæ live in the leafbuds at the tip of the branches of the dwarf willows which fringe the pools and river-sides in every part of the country, and there form rosette-like galls.

The wart-like galls common on the meadow-sweet (Spiræa) ; the uneven swellings on the stalks and leaves of the stinging-nettle ; the little furry. purses on the ground ivy; the woody, shapeless excrescences on the raspberry plants; the slender upright growth on beech leaves ; the blisters on bedstraw, yellow nettle, and others; the knots within the very blossoms of many flowers ;—all these, and a great many more, are the work of little creatures in this numerous family.

A few of them have been found in woody galls of Hymenopterous insects ; some others are Aphis-eaters.

Besides these, some of the larvæ live within the stalks of groundsel and other allied plants ; others live in rolled leaves, and one, the *Cecidomyia tritici*, is a scourge to the farmers, inhabiting the ears of corn. It will give some idea of the destruction of which these insects are capable, to relate that forty-one of their maggots have been counted in the husk of a single grain of wheat! The Hessian Fly of America belongs to this family.

Some of the larvæ in Cecidomyzidæ spin silken cocoons.

3. *Bibionidæ.*—This is a small family of rather stouter

and less Gnat-like insects than the preceding. Their larvæ feed, some on the roots of grass, others on dung and dead animal or vegetable matter of various kinds. The Flies have generally three ocelli.

In the next three families the larvæ are chiefly aquatic, and the perfect insects blood-suckers.

4. *Simulidæ.*—This family contains only one genus, and that but few British species—perhaps we should be as well satisfied if we might also say but few individuals, for they are a race of tormentors. They are the *Sand-flies* of Northern latitudes, and the *Mosquitoes* of some countries, while in others the name Mosquito is given to various species of Culex. In the preparatory states they are aquatic, the larva spinning a cocoon for the purpose, which, however, is but partially enclosed, the fore-part of the pupa being subject to the action of the water. This family is distinguished by the tibia and the first joint of the tarsus being somewhat broad and flattened.

5. *Chironomidæ* are the Midges—an immense family of beautiful but bloodthirsty little creatures. A little, thin, wriggling, red, eel-like Maggot, common in stagnant water, and known as the Bloodworm, is the larva of a common species, *Chironomus plumosus*, with feathery antennæ.* Like the pupæ of the Culicidæ, or true Gnats, the aquatic pupæ in this family are capable of motion, though less active than the Culicidæ. They live at the bottom of the water, rising to the surface when about to change. Some of the larvæ form tubes of

---

* It is remarked that the feathered antennæ, so beautiful (especially in the male) in many of the Nemocera, belong only (or generally) to species with aquatic larvæ.

decayed leaves spun together by silken threads, in which they live.

The family contains one, probably more, species living in dung in the earlier stages.

The legs of these insects are slender and not flattened.

6. *Culicidæ*, the true Gnats, are, some of them, so well known as to be recognisable at once. Their size is generally greater than that of insects in the preceding families, and they are distinguished by the position of the long proboscis, which is held projecting straight forward (fig. 70, p. 340). This organ, comparatively inconspicuous in the female, is often exceedingly ornamental in the male; its feathered palpi, combined with the feathered antennæ, forming a most beautiful cluster of plumes upon the head (Pl. XIV., fig. 1). But, while the apparently simple, needle-like proboscis of the female (see fig. 70, p. 340, and Pl. XIV., fig. 1, b) is less likely to attract the eye than are the showy plumes of her more warlike-looking mate, hers is the weapon with which we are but too well acquainted in action. This little needle, finer than any hair, consists of a long tubular sheath, which, enclosing and guarding five minute lancets, serves also as a sucker to draw up blood from the vessels pierced by these instruments. These delicate but efficient little serrated lancets are either wanting in the male, or are much less fully developed, and the female alone is bloodthirsty. So also is she alone musical, and musical only at her most bloodthirsty times; and the trumpet which we hear in the dead of night sounding the attack is the instrument of this amazon, and seldom gives a false alarm.

The swelling and irritation which follow the bite of a Gnat are accounted for by supposing that she injects a venomous fluid into the wound, which creates inflammation, while the swelling is caused by a pouring of fluid from the vessels into the tissues around, an effort, probably, to free the blood from the poison, but also having the effect of rendering the little blood-sucker's draught easy and more copious. It has long been a fact familiar to many people that the bite of a Gnat which is allowed to suck its fill, is much less troublesome afterwards than that of an insect disturbed while sucking. Humboldt has stated this of a South American Gnat, and is quoted by a writer in the "Zoologist," who had found the same thing true in England. In a paragraph recently extracted by the *Times* from an American paper, a gentleman who has lived for years in a Mosquito country, states the same fact, and accounts for it by saying that the Gnat having leave to drink his fill, sucks back the poison together with the blood.

The lively Gnat larvæ are well known, as, with the big-headed, tadpolish, jerky pupæ, they are to be found in abundance in all standing water, not only in ponds and ditches, but in cisterns and tanks, and occasionally in our very water-jugs.

Yet tenfold the number of larvæ which our eyes have ever looked upon would not seem enough to account for the cloud-like myriads of Gnats which are often seen filling the air. In the "Insect Miscellanies" a swarm is recorded so dense as to have appeared like smoke issuing from the spire of Salisbury Cathedral, giving rise to an alarm of fire. It is not possible to conceive the immense number of such minute creatures which must be congregated together to become visible at so great a height.

The flight of these insects is worth noticing, when we consider the astonishing muscular power which must be exerted by animals of weight so inconsiderable, in maintaining their position against the wind. They generally fly, or hover,* with their heads towards the wind, and a cluster may be seen for hours dancing in the air without yielding one inch of—we cannot say—*ground!* The loud humming, or trumpet-like sound too of the female, if, as seems most probable, produced by the rapid vibrations of the wings, must require a marvellous array of powerful muscles.

That this is indeed the case is easily to be seen in a Gnat rendered transparent by soaking in turpentine and then viewed by polarized light. The whole of the bulky (!) thorax (the only part of a Gnat which seems to have any solidity at all) appears crossed and re-crossed at right angles by broad, band-like muscles, which (if the selenite crystal be used in the examination), actually gives the little creature the appearance of being dressed in a large check tartan jacket! A beautiful economy of power is also to be noticed in the centralization of the weight of the insect in this one part, by which the poising of the body is effected without special muscular effort.

Rapidity of motion in these insects is in evident response to quickness of vision. Though doubts are entertained as to the *distinctness* of sight in insects generally, there can be none concerning the swiftness

---

* The verb *to hover* is used here and elsewhere to express the action of flying, fluttering, or remaining as it were suspended *over one spot*, whether this be as, when by the rapid action of the wings the insect is maintained apparently motionless in the air (as in some Syrphidæ, &c.), or as in the limited dance of the Gnats.

with which they perceive and avoid danger. Who has not watched a company of these creatures maintaining their merry dance during a shower of rain, without a single individual being caught and dashed to the earth? It is clear that, as has been said of certain Tipulæ also, they *dodge* the drops of rain; and we may conceive this necessity as adding greatly to the glee and excitement of their gambols. How clumsy and tame a performance is the boasted "sword dance" in comparison with this!

7. PHLEBOTOMIDÆ, notwithstanding the threatening name, and (8) HETEROCLITÆ, appear to partake of some characters both of the Gnat and the Daddy Longlegs. Some are blood-suckers and others not, and the food and habits of the larvæ are various. The genus generally best known is Psychoda (in family 7), which contains only two species. One of these, Psychoda phalenoides (or "Moth-like" Psychoda), a little harmless grey woolly-winged insect, sometimes libellously called a Midge, is commonly to be found on the window-panes and elsewhere in-doors, especially in winter. It is less than one-twelfth of an inch long, with broad wings, sloping, roof-like, from the back and antennæ, which are banded with black and decorated with whorls of hair, and exceedingly beautiful, though not more so than hundreds of other species in Nemocera. The Fly runs actively, but is more noticeable for its habit of making sudden hops (produced, however, by wings, not legs) in all unexpected directions. The larva is terrestrial, and lives in dung.

9. *Tipulidæ*.—This family contains the "Daddy Longlegs," "Harry Longlegs," or "Craneflies," the largest insects not only in Nemocera, but if *length*—length of body and length of limb—is considered, the largest of English Diptera. The Tipulidæ are easily to be distin-

DIPTERA.—NEMOCERA. 351

guished from the other families in the group by a transverse seam in the thorax, which is not found in these.

The body of the female terminates in a sharp-pointed oviduct; that of the male is abruptly truncated. These insects, while out of their proper place they are among the most awkward of animals, are interesting as examples of structure peculiarly fitted for an especial purpose. Everywhere but at home the Daddy Longlegs is so encumbered

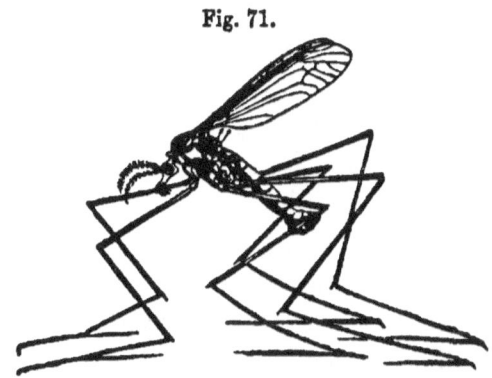

Fig. 71.

Daddy Longlegs (*Tipula oleracea*).

by his long and unmanageable limbs, that his life seems to be spent in clumsy but successful efforts to leave them behind him on every possible occasion and in every possible place. But the Daddy Longlegs *at home* is another creature, and is conspicuous as an example of a very singular mode of locomotion. The larvæ are subterranean, feeding on the roots of grass; the female, therefore, seeks grassy meadows where to lay her eggs. Her progress among the long blades is quick and easy, the slender-jointed legs curling round and embracing these leaves, so that, except for the body being maintained in a standing rather than a hanging position, the action might be compared to that of the Sloths, or Tree Monkeys, with their prehensile limbs and tail.

The destruction occasioned by the larvæ is sometimes very great, though at others the quality of the grass is even improved by the eating away of superabundant roots. About ten or twelve years ago the insect swarmed in and near London; in Kensington Gardens, and other places, large patches of ground were entirely denuded.

The long-nosed brown pupa-case of the Daddy Longlegs may easily be found, empty, and projecting from the turf, looking like a legless, wingless skin of the perfect Fly.

The head of this insect may be recommended to young microscopists as very easy of preparation; and, from the beauty of the whorled antennæ, the large size of the compound eyes, and the easily displayed structure of the mouth, is a very interesting object.

In the Tipulidæ family the larvæ are chiefly subterranean, feeding on fungi, roots, &c.; but among them are some which are aquatic in the larva state, and bloodsucking in the perfect.

10. *Rhyphidæ.* There are two species of this family, both common, and though smaller and shorter-bodied in proportion than those of the last family, they resemble them more nearly than the preceding. They are, however, at once distinguished by the absence of the transverse suture down the back. The larvæ are dung-eaters.

Many of the flies in Nemocera, as the Gnats, Craneflies, &c., fly both by day and night. The Flies of the following division, Brachycera, are all diurnal.

CHAPTER XXVII.

DIPTERA.—(*continued*).

BRACHYCERA, the second large section of the Proboscidæ, contains seventeen families :—
1. Stratiomidæ (*Soldier-flies*).
2. Xylophagidæ.
3. Tabanidæ (*Horse-flies*).
4. Acroceridæ.
5. Asilidæ.
6. Leptidæ.
7. Bombylidæ (*Bee-flies*).
8. Scenopinidæ.
9. Empidæ (*Snipe-flies*).
10. Dolichopidæ.
11. Lonchopteridæ.
12. Platypezidæ.
13. Pipunculidæ.
14. Syrphidæ (*Dragon-flies, &c.*).
15. Conopidæ.
16. Muscidæ (*Houseflies, &c.*).
17. Œstridæ (*Gadflies*).

These families form three groups, distinguished chiefly by the character of the metamorphoses.

In the two first the pupa remains within the skin of the larva, which retains something of its original form.

In the eleven following families the pupa sheds the larva skin.

In the four last families the pupa remains within the larva skin, which shrinks and hardens into an even, somewhat eggshell-like or cocoon-like covering for the pupa.

The antennæ in the first two groups are generally long in proportion to their bulk (Pl. XIV., figs. 2 a, 4 a); or, if short and thick, have a *terminal* bristle (figs. 3 a, 6 a). In the third group they are generally short, and have the last joint much the most bulky, and garnished with a bristle springing from its upper side (Pl. XV., fig. 4 a; XVI., figs. 3 a, 5 a).

The family Stratiomidæ contains the Soldier-flies (Pl. XIV., fig. 2, *Stratiomys chameleon*), large or middle-sized and somewhat broad-bodied insects, very prettily marked in spots, streaks, and triangles, and of bright colours—black variegated with red, yellow, orange, or white. The hinder part of the thorax is generally armed with spines; in the genus Stratiomys there are two; in other genera the number varies—one, Beris, having in some species as many as eight. The feet have three pads.

The larvæ in this family are chiefly aquatic—some, however, being terrestrial and feeding on decomposing matter. The perfect insects haunt flowers, of which they suck the juices. In some genera, as Sargus, the Flies are more slender in form, and of beautiful metallic colouring. *Sargus cuprarius* (Pl. XIV., fig. 3) is a common and beautiful insect, with golden-green head and thorax, and an abdomen lustrous with the purple and gold of the Copper ore called "Peacock." The larvæ of these live in the earth.

Xylophagus, in the next family, Xylophagidæ, contains but few and rare species, which in the earlier stages live in decayed wood.

The antennæ in these two families are composed of from five to ten joints, and are generally somewhat elongated, but less distinctly articulated than in the Nemocera section. In Sargus and some others of the Stratiomidæ, the third and following joints are consolidated into a sub-globular, oval, or spindle-shaped mass, in which the articulations are to be distinguished, and which is terminated by a bristle. The antennæ in *Stratiomys* are long and elbowed (Pl. XIV., fig. 2 *a*).

Tabanus is the principal genus in the third family, Tabanidæ (Pl. XIV., fig. 4). The "Horse-stinger," a speckled grey Fly, about half an inch long and very common in woods, is an insect well known, especially by those who have once felt its peculiarly acute bite. There are several species, of which one, the Oxfly (*T. bovinus*), is nearly an inch in length, broad in proportion, and covered with a handsome chequered pattern in grey and white. These insects are called Horseflies and Oxflies, Clegs, and Gadflies (a name more usually applied, however, to the Œstrus or Botfly), and, from the sound produced by their wings, Breezeflies. In some countries they are a scourge to the cattle, and even in England the approach of a single individual will sometimes occasion no small panic in a herd. With these insects, however, as with the Gnats, blood-sucking appears to be the habit of the female alone, the males being true flower-lovers.

The antenna consists of three distinct portions—viz., a basal joint, a second, and a tapering mass composed of several joints.

The eyes are often exceedingly beautiful, lustrous and

varied in colouring. Some are golden green, with lines and spots of purple, others are bronzed, others purplish green, others green-striped or banded with crimson.

The feet have three pads.

The larvæ are terrestrial.

Acroceridæ, the fourth family, contains a few species of small but remarkably swollen-looking Flies, with nearly globular abdomen, broad, very high, and convex thorax, and exceedingly small head, which seems only just large enough to contain the eyes. The antennæ are small, and placed close together; the feet, like those of the Tabani, have three pads.

These insects are sluggish, and haunt flowers.

Asilus, the principal genus in the family Asilidæ, contains some long, strong, hairy, hungry-looking Flies of predaceous habits. They feed principally on other insects, and an Asilus may often be seen on the wing bearing the whole weight of an unlucky Bee—probably on his way to a picnic, as there seems no other way of accounting for his not eating his dinner where he found it. It is supposed that some species suck the blood of quadrupeds. *Asilus crabroniformis* (Pl. XIV., fig. 5) is a very handsome and conspicuous Fly, clothed all over with deep golden hairs, except on the fore-half of the abdomen, which is velvety black. The wings are somewhat golden or tawny, with dusky patches on the hind margin.

The larvæ of the Asilidæ live in the earth or in decayed wood.

In the Leptidæ the antennæ are nearly all very short, being composed of three short joints (of which the last is generally the greatest in circumference), and a long terminal bristle usually reckoned as a fourth joint. The feet in Leptis have three pads, as in some other predaceous Flies already described. *Leptis scolopacea* (Pl. XIV., fig. 6) is a Fly often found in woods, lanes, and gardens. It is of a somewhat slender and tapering figure, and of much feebler aspect than the Asilidæ. The thorax is of a pale striped grey in the male, yellowish in the female, the abdomen tawny-coloured, with a row of black spots running down it, and the grey-tipped wings are spotted and partly bordered with brown. The larvæ live in the earth or in decayed wood, and that of one foreign species is said to catch the small insects upon which it preys in pitfalls formed in the sand, somewhat after the manner of the Cicindela larva. Common as are instances of such instinct among the larvæ of other orders, it is rare to find them in the Diptera, cocoon-spinning being almost the only constructive work performed by them, and that being confined to a few families. One common species in this family, *Atherix ibis*, a little ash-coloured (♀) or tawny (♂) Fly, spotted and banded with black, and about a quarter of an inch in length, with an aquatic larva, has the curious instinct to lay its eggs on branches overhanging the water, into which the larvæ falls on emerging from the egg.

In Bombylidæ, the most conspicuous is the Bee-fly, *Bombylius major* (Pl. XV., fig. 1). This is a furry-looking Fly, with a small head, wide thorax, and abdomen wider still and somewhat flattened. It is black, covered with bristly golden hairs above, and with black and white hairs below. The wings, even when in repose, are kept at full stretch, and are longitudinally divided in colour, the costal half being dark brown, the other clear and colourless. The most remarkable feature, however, is the long, slender, projecting proboscis, little inferior in length to the body of the insect. The antennæ are long, pointed, and slender.

Like the Humming-bird Moth, the Bee-fly may frequently be observed eagerly hovering over a flower, and, without settling, extracting honey by means of its long proboscis. It resembles the Moth too in the suddenness with which it darts away if approached while thus suspended, apparently without motion, in the air.

The history of its larvæ is not yet fully known, and it is a question whether they are parasitic in their habits, or vegetarian. There are some insects in the family which are known to be parasitic, but most of these are rare. Thereva, one of the genera, contains predaceous Flies.

The eighth family, Scenopinidæ, contains a few common but inconspicuous little black insects with rust-coloured legs, found in houses and stables.

Empidæ is a numerous family of voracious insect-eating Flies somewhat resembling Asilus in figure, but generally of small or middle size. Their heads are small and round, the antennæ three-jointed, tapering, and terminating in a point or bristle of one or two joints; the tongue is generally long, and when at work very conspicuous, whence these insects are sometimes called Snipe-flies. The wings are without the rounded lobe in the axil, which appears in nearly all the other Brachycerous flies excepting those of the families Dolichopidæ and Lonchopteridæ.

Empis tessellata (Pl. XV., fig. 2) is a dull, ash-coloured Fly, with three longitudinal black streaks on the thorax, legs long, strong, and spined, wings dusky and tinged on the fore-part with reddish brown. It is very common in the spring, and may be met with on any plant frequented by other insects.

Another insect, not common, but never to be forgotten when once seen, even though but slightly magnified, is the *Rhamphomyia pennata*, a little black Fly, the male of which is rendered noticeable even to the naked eye by the apparent thickness of the legs. These on being magnified are seen to be most beautifully feathered, or fringed throughout their whole length, by broad, flat, striated scales or hairs. The decoration of the female is much more insignificant.

Another, and much more common Fly in this family is the little *Hilara*, the males of which may often be recognised by the inflated appearance of the first joint of the tarsus (metatarsus), which is nearly globular. This insect abounds in gardens.

Some of the Empidæ are found hovering in swarms, like Gnats, over water. The family contains many genera and many species.

Dolichopidæ also, the tenth family of Brachycera, contains a large number of genera and species. The Dolichopidæ are small bright Flies, generally with a metallic lustre and colouring, very brisk and joyous in their movements, especially in running, and, like most very active insects, predaceous in their habits. Many species in this family are found near water, and even on its surface, upon which they run actively in search of prey. The larvæ are, however, supposed to be terrestrial, and it is probably the insects abundant in such situations which make them attractive to the Flies.

The remaining three families of Flies with naked pupæ contain few species, and their habits are not very remarkable.

Of the four families in which the pupæ remain within the altered and cocoon-like larva skin, the first is Syrphidæ.

This is a very large family, consisting of no less than thirty-one genera, and containing many large Flies, rendered conspicuous by their bright and well-marked colouring, their vigorous flight, and their constant presence during spring, summer, and autumn. Not only in fields, lanes, and woods, but in our gardens too, the motion of their bright and glancing wings, their musical hum, and their evident enjoyment of the sunshine and the flowers, are no small ingredients in the general brightness.

They are, almost without exception, pure flower-lovers in their perfect state. In their earlier stages some of them are useful servants to the gardeners, for while a few, feeding on bulbous roots, &c., are known as injurious to cultivated plants, the larvæ of the genus *Syrphus* and others are highly beneficial to them, living entirely upon Aphides. Some, less praiseworthy, are parasitic in the nests of Humble Bees and of Wasps. Others live in rotten wood, cow-dung, and fungi; while the larvæ in the genera Eristalis and Helophilus are aquatic.

The "Drone-fly," of which the aquatic larva is described p. 47, is *Eristalis tenax* (Pl. XV., fig. 3). This, and some other large stout-bodied Flies in this genus and Helophilus,* must be familiar to every lover of the garden from their habit of holding revels in the blossoms of the Michaelmas daisy. They make a peculiarly loud and musical hum on

---

* These two genera much resemble each other, and may be known by the form of the subapical cell, into the middle of which (see fig. 22, p. 49) the cubital nerve makes a sudden dip. In Eristalis the *subcostal and radial nerves meet* before reaching the margin; in Helophilus ending separately in the margin.

rising from the flowers if disturbed. The pupa of Eristalis tenax, in and out of the puparium or larva skin, is figured p. 5, and the reader will observe that the long tail, used by the larva as a breathing tube, still remains attached after its change.

Another Fly, not occurring so abundantly, yet common enough and remarkable enough to attract attention, is *Volucella pellucens*. This insect is noteworthy both as one of the Parasitic Flies and also on account of its very singular appearance, which arises from the perfect translucency (and semi-transparency) of the basal half of the large oval abdomen.

The thorax and the hinder-half of the head are black, and the effect of the clear, colourless, and apparently empty fore-half of the abdomen, is heightened by the colouring of the wings, which, clear at the base, have a brown mark across them exactly corresponding with the darkened half of the abdomen. The antennæ are beautifully feathered. Another species, V. Bombylans, is hairy, and has a remarkable resemblance to the Humble Bees, whose nests it enters for the purpose of depositing its eggs.

The genus *Syrphus* contains some Flies similar in form to the preceding, and others of smaller size with narrow linear abdomens, and always marked with black and yellow, or whitish bands or spots. They have a peculiar mode of hovering, apparently motionless, over flowers, making sudden darts forwards and from side to side. If, however, the apparently motionless Fly be observed with attention, it will be seen that the rapidity of motion in the wings is so great as to render them almost invisible. Syrphus pyrastri, a common species, is figured Pl. XV., fig. 4. The long pear-shaped or flask-shaped brown pupa case of this insect may be frequently found attached to the leaves which the larva (whose skin now forms this pupa case) has but lately cleared of the infesting Aphides. The larva is a slimy, whitish, slug-like, or rather *leech-like* grub, with a curious habit of seizing its victim and holding it raised in the air until all its juices are extracted.

The many genera which this family contains present numberless varieties in form, colours, and marking. Thus, as described above, some are Bee-like in form, others compact, with oval or nearly globular abdomen, others are long and narrow, and others again have a somewhat club-shaped abdomen (see Pl. XV., fig. 5, Melithneptus menastri). In some of the more slender species, the thighs are swollen, giving the Fly the appearance of a leaping insect; others have peculiar tarsi. The colours also vary: they are black and yellow, black and dull red, black and white, black and grey. Some species are black only, others lustrous and metallic, while others have black and green, yellow, or metallic and glittering heads with black and yellow banded bodies.

Notwithstanding all these variations in form and colour, there are characters which make it easy to distinguish the Syrphidæ. The head is convex in front and flat behind, so as to be *nearly all face*—a large round face almost covered by the eyes, which are especially large in the males. The tongue, which is large and well developed,* is bent about the middle when at rest, and the front of the head sometimes forms a sort of projecting beak or snout (very conspicuous in a large, dull, red Fly, the *Rhingia rostrata*. The antennæ are, almost

---

* The tongue of the larger Syrphidæ is one of the most beautiful and most easily prepared for the microscope, of the tongues of the Diptera.

without exception, of three joints, of which the last is large, nearly globular, oblong-oval, or somewhat kidney-shaped (Pl. XV., 4 a, Ant. of Syrphus pyrastri), and bears a curved bristle, which is often beautifully feathered. The wings afford the best character, the *subapical cells* being perfect—*i.e.*, bounded short of the margin by a transverse vein. There are also two false veins, one of which is conspicuous and runs from the prebrachial into the subapical cell (see fig. 22, p. 49).

The Conopidæ are a small family of prettily-coloured, but rather awkwardly-formed Flies, of which the larvæ are parasitic in the bodies of Bees. They are rather slender, generally about half an inch in length, with the abdomen thickest toward the end, and curved downward. The tongue is long, stiff, and projecting; the antennæ are long and of singular form; three principal joints, of which the second is the longest and the third the thickest, are terminated by three small joints, the last and the last but one forming a double joint to the antennæ. The wings have the perfect subapical cell of the Syrphidæ, but not the false veins. There are two species common—*Conops quadrifasciatus*, a black and yellow Fly, with the abdomen not remarkably small at the base, and *C. rufipes* (Pl. XV., fig. 6), in which the abdomen is, as it were, set on a stalk composed of the attenuated basal joints.

We have now arrived at the Muscidæ, by far the largest family, and one which, as the Linnæan genus *Musca*, included many of those now distributed among other families.

The reader will at once recognise some familiar insects in this family, as the Housefly, the Bluebottle, the chequered Blowfly, and the common yellow Dungfly.

The thick, short tongue of the Housefly (*Musca domestica*, Pl. XVI., fig. 3), with its large two-lobed extremity, and capability of being drawn entirely into the mouth when not in use is well known. The antennæ somewhat resemble those of the Syrphidæ, and the wings are sparsely veined, and sometimes nearly or quite without the little alulæ or winglets.

This last peculiarity is used to subdivide the family into the *Calypteræ*, in which the alulæ are large, and the *Acalypteræ*, in which they are wanting, or very minute.

In the habits of both larva and fly in the several groups of genera in this one family, there is nearly as much variety as in all the other families of Brachycera together. Thus, while among the Flies are found flower-lovers feeding on honey, blood-suckers, Flies preying on others, Flies oviparous, and Flies ovoviviparous; among the larvæ are found some terrestrial and some aquatic, carrion-feeders, vegetarians, and wine-bibbers; parasites in the nests, and parasites in the bodies of other insects; gall-makers, and leaf-miners. Of the Flies with parasitic larvæ there is a very large group, of which one genus alone, *Tachines*, contains more than one hundred and sixty species, varying in length from one-eighth to two-thirds of an inch. They are powerful Flies, and some are of brilliant and metallic colours, while others are dull-coloured, hairy, and unattractive. Nearly all the species are rare. The commonest Fly in the group is *Bucentes geniculatus* (Pl. XVI., fig. 1), a blood-sucking, dowdy-looking Fly, frequently found in the house, and which is remarkable as having the proboscis doubled under, about the middle, as in the Syrphidæ.

Another insect with the same propensity to "bite" men and beasts, is *Stomoxys calcitrans* (Pl. XVI., fig. 2), the Stable Fly. Bred in dung, and very common in stables, it is unfortunately not rare in houses, where its bloodthirsty habits bring discredit on the harmless little Housefly (Pl. XVI., fig. 3), which it closely resembles, but from which it may readily be distinguished by its *slender proboscis, which projects in front of the head.* The proboscis is geniculated near the base. It is, indeed, a near relation, being in the same group as the genus Musca, to which the Housefly belongs; but whereas, in the genus Stomoxys, a *projecting* slender, polished, and needle-like proboscis forms an admirable instrument of torture; in Musca the soft, short, fleshy tongue, with the sight of which we are all so familiar, is totally incapable of wounding the skin. The tongue of the Housefly is indeed adapted only for licking up such fluid substances as are entirely unprotected and left exposed to its action. The little creature has, however, a ready mode of rendering soluble substances fluid, by emitting a drop of clear water from the mouth from time to time. Hard white sugar and similar substances by this means become fit food for the tender little mouth.

The larvæ of the Housefly, like those of Stomoxys, are found in dung.

The " Bluebottle" Flies, so well known and so little loved, are also species of the genus Musca. Of the *habitat* of their larvæ the reader needs not to be informed. The " Greenbottle," a smaller, brighter, and prettier Fly, is of the same genus. These Meat-Flies, as they are called, are resembled in their habits by the Flies of another group, also flesh-eaters, and thence called *Sarcophaga.* The large, handsome chequered Blowfly is one of these, and is worthy of remark as being viviparous (or, more correctly, ovoviviparous), a peculiar provision, of which the evident purpose is the saving of time in the removal of dead matter. Réaumur calculated the number of young produced by one Fly of this species to be about 20,000.

From these we turn to the so-called Flower-flies, or Anthomyia, the larvæ of which live chiefly on decaying animal or vegetable matter, in roots, as onions, radishes, &c. The perfect insects are found on flowers.

They are generally dull-coloured, hairy Flies, of various shapes, some being short and thickset, and others of a more slender form. Some have the abdomen spotted or chequered, others are black with grey or greyish-white hairs, tawny, or pale dull red, and sometimes aeneus. This genus contains a large number of species, of which one, *Anthomyia betæ*, mines the leaves of mangold wurtzel, occasioning great loss to the farmers.

All the Muscidæ hitherto mentioned belong to the first division of the family, and have large alulæ.

The second division, *Acalypteræ*, contains Flies in which the alulæ are wanting, and which are generally smaller and of lighter make than the large Muscæ, &c., lately described.

The clean, bright, active *Scatophaga*, or yellow Dung-fly, is one of these. This pretty little Fly is familiar to every stroller in the country, and is no less frequently met with basking singly in the sunshine in the flower-cups than clustering with many others on the unclean mass whence the approaching footstep drives it with a mighty buzz and bustle.

The egg of the Scatophaga is a beautiful and curious object. It is a long curved egg (about half an inch in length), convex in front and nearly straight behind, with the top cut off obliquely from front to back downwards, and covered by a hinged-on lid, which completely closes the opening. But the appendage which most of all strikes us as evidently adapted to a certain end, is a pair of long arm-like proboscs which arise from the upper end of the shell, and spreading out (as a man sinking in mud would spread out his arms), prevent the half-buried egg from sinking entirely in the soft mass of dung. Hundreds of these little white specks may be seen in a mass of cow-dung on a summer's day, and are well worth examination. The surface appears covered with a network.

The perfect insect is predaceous, and frequents flowers probably for the same reason as do other predaceous Flies, partly perhaps for the sake of their honey, but chiefly on account of their being the resort of other insects. The foot of this Fly is a very beautiful and very easily prepared microscopic object; the pads are long and somewhat pointed, and covered with unusually long hairs or suckers.

The larvæ of others of the Muscidæ are Leaf-Miners, like certain Lepidoptera—the Honeysuckle, Holly, and Columbine being conspicuously attacked by them; others again, as has been said, form galls in various parts of plants, especially of the Syngenesia, as the Thistle and others.

The genus *Tephritis*, or *Trypeta*, is conspicuous among these. Most of the species are very small, but some are more than a quarter of an inch long. They may generally be recognised by the brilliancy and beautiful colouring of their eyes, and by their delicately-painted wings, which are brown, grey, or black spotted, and banded in various patterns.

The genus *Chlorops* is especially mischievous in its larva state to grain of various kinds. One species, the little Chlorops lineata (or *Striped Green-eyes*) is one of the commonest, and is frequently to be met with in houses, sometimes in large numbers. It is yellow, with five black stripes on the thorax, and a black spot on the abdomen. The ravages committed by this insect and its relations, both in England and in other countries, are in some years of great importance, whole crops becoming " gouty " under their attacks.

In the genus *Drosophila* is found one insect of decided anti-teetotal habits—a sad set-off to the good little stories of good little dogs who sit outside public-houses while their masters are sitting inside. This melancholy example of a beast which is no better than a man, lives in its larval state in the casks containing fermented liquors, feeding on the rich substances deposited there. *Faute de mieux*, it will feed on fungi or oakapples. Others of the genus are leaf-miners in chickweed, catchfly, corncockle, peas and other Papilionaceæ, and on some of the Cruciferæ.

Nearly related to these is the little black Fly proceeding from the hopping maggots common in cheese, with another found in bacon, probably the most active of any terrestrial Dipterous larva. The maggot effects its spring by first standing on its tail, then, curving itself into a circle and grasping its tail with its jaws, " it next contracts its body into an oblong, so that the two halves are parallel to each other ; it then lets go its hold with so violent a jerk that the sound produced by its mandibles may be readily heard, and the leap takes place."

The last of the Muscidæ which shall be mentioned here is a little dark, shining, metallic-coloured Fly, less than one-sixth of an inch in length, with clear wings, marked near the tip with one black spot. This Fly, *Sepsis cynipsea* (Pl. XVI., fig. 4), can hardly fail to be observed, owing both to its great frequency and also to the peculiarity of its manner and motions as it runs actively over the surface of laurel and other leaves, with its glittering little wings raised almost perpendicularly from its body, as if in act to fly. The larva is a Scavenger.

Œstridæ, the last family of the Brachycera, is that which contains the insects well known in their larval stage under the name of *Bots, Wurmals*, &c. The perfect insects, or Bot-flies, are also called Gadflies, in common with the Tabani, from which nevertheless they differ in important particulars. If, however, either of the etymologies be just, whether that which assumes the Gadfly to be the *Goadfly*, goading the cattle to almost a state of madness, or that which considers it as the Fly which "makes the cattle gadde up and downe with stinging them," both Tabanus and Œstrus are fully entitled to the name. On the approach of either, the herd of cattle, oxen, or deer, or the flock of sheep, is thrown into the utmost terror and dismay, oxen "gadding up and downe," while sheep herd together in their sheepish belief that in numbers there is safety. While, however, the attacks of the bloodthirsty Tabanus arise from purely selfish motives, the Œstrus approaches the terrified victim with another end in view. Another end it is in more senses than one, for no blood-sucking proboscis is hers—while Tabanus is fully armed with lancets and sucker, her mouth is in a merely rudimentary state, and altogether incapable of aggressive operations. Not so her long and sharp and jointed, possibly also poison-dropping *ovipositor*. This is her goad, this her weapon of attack—and not care for herself, but care for her progeny, the motive of her approach.

The Botflies all deposit their eggs on or in the bodies of various quadrupeds, some species choosing one part and some another, but each species of Fly being constant both to the one species of quadruped and to one part of its body.

The Botflies are large and hairy-bodied, and carry their wings extended. Œstrus Bovis, the Ox Botfly, is about half an inch in length; the legs are long and strong, and the alulæ very large. It is of a blackish brown colour, banded with black and coloured hairs. On the face the hairs are reddish, and pale yellowish on the head. The thorax is streaked with reddish hairs, and the base of the abdomen is clothed with the same colour, the tip being orange red.

The habit of this insect is to deposit its eggs, singly, in holes which it perforates, with an auger-like ovipositor, in the backs of oxen and other horned cattle. The egg thus placed causes a large open tumour, in which the larva resides until ready for the pupa change, when it emerges, falls to the ground, and there undergoes its metamorphosis.

The Botfly of the sheep, *Cephalemyia ovis* (Pl. XVI., fig. 5, 5 a), lays its eggs in the nostrils of its victim, whence the larvæ creep farther into the head, sometimes with fatal effect upon the animal. It is to avoid the attacks of this Fly that sheep may be seen holding their noses in the dust, in ruts, or dusty places, or stamping and shaking their heads, or running violently about the field. This Fly is about the same length as the former, but the legs are much smaller and less powerful. It is of a dark colour, chequered with hoary hairs.

The horse is subject to the attacks of four species of Botfly, of which, however, only one is common. This insect, *Gasterophilus equi*, deposits its eggs, not in their final place of rest, which, being the stomach of the horse, is inaccessible to the Fly, but on the surface of its body, *within reach of the tongue*. The animal, by licking itself and swallowing them, transfers the unwelcome little stranger into its own inside, where it grows and prospers, till, as in the former cases, it detaches itself, and discharged by the horse, falls to the ground, and undergoes the final changes. This Fly is sometimes as much as two-thirds of an inch in length. It differs from both the preceding in the very small size of the winglets. The legs are of moderate size and strength. It is somewhat rust-coloured, black at the tip of the abdomen. There are varieties, differing in colour.

Another species of Gasterophilus lays its eggs on the lip of the horse.

Besides the above-mentioned Flies, there are some rarer species infesting deer; and in other countries, several more are known, one even selecting the rhinoceros as its host, while others have been known to attack men.

The third sub-section of the Proboscidea is *Hypocera*. The antennæ themselves differ little in this from some of the Brachycera, but are differently placed, being very near the mouth, instead of as in the Brachycera, between the eyes.

This division consists of but one genus, Phora, containing only a few species of minute Flies, ranging in size from one twenty-fourth of an inch to one-sixth of an inch. Some of these are exceedingly common, and haunt the window-panes throughout the year. They are rather hump-backed little creatures, with deflexed, fringed wings, entirely destitute of transverse veins, and move with much activity. In the larval stage, some species feed on decayed vegetable matter, while others are supposed to be parasitic on other insects.

The characters of the second section of Diptera, EPROBOSCIDEA, have been given at page 339.

The Eproboscidea are, as has been already said, all parasitic, in the perfect state, in various birds and quadrupeds. The Forest-fly, or *Hippobosca*, is a well-known example, as also the stout-bodied, wingless *Melophagus ovinus* (Pl. XVI., fig. 6), or "Sheep-tick," as it is improperly called, remarkable as having the appearance of two groups of simple, rather than a pair of compound eyes. This is owing to the external faces being distinct from each other, and round instead of hexagonal. These insects are parasitic on horses, oxen, and sheep. Others of the same family are parasitic upon birds, and the swallow is especially subject to their attacks.

An unusual circumstance occurs in this family, the female giving birth to but one individual, and that not until it has either already attained the last stage of larvahood or has become a pupa.

The second family in this section contains only two known species, parasitic upon bats, and very rare. They are wingless, and of most singular appearance, the head being thrown completely backwards, and carried in an inverted position. They form the genus *Nycteribia*.

With the Diptera, the most modern arrangements of insects come to a close. There remain, however, a few genera whose claim to rank as insects has so often been urged, and whose appearance is so insect-like, that some mention of them seems desirable. These are the *six-legged*, non-changing Lepismas, the Spring-tails, the Lice, and the Bird-lice. These have been variously arranged; at one time with the eight-legged Acari, &c., among the Arachnida at another, with the Fleas in the order " Aptera," and so on.

Whilst, however, the Flea, although apterous, is clearly shown by the nature of its transformations to belong to the true insects, the Spring-tails, Lice, &c., are excluded from this class by the total absence of transformations, and have been formed into a class by themselves under the name Ametabola.

This, again, is divided into Thysanura and Anoplura.

To the first belongs the little silvery fish-like *Lepisma*, whose delicate scales are well known to microscopists as a low test object. The abdomen is furnished with several bristle-like tails, of which three are the most conspicuous; the mouth is mandibulate, the antennæ are long and bristle-like, and without the tail it measures about one-third of an inch.

The Podura, or Spring-tail, in the same section, is a very curious little creature, effecting its spring by means of a forked tail turned under the abdomen, and acting precisely like the wood-and-catgut spring of the wooden frog made for children. It is smaller than the Lepisma, dark and velvety, with the thorax and abdomen tolerably distinct, and has large, thick antennæ and mandibles.

These insects are found among wood, in the sawdust in cellars, under stones, &c.

Anoplura contains the Sucking Lice, *Pediculi*, and the *Nirmi*, or Biting Lice, which, except one species which infests dogs, seem to be confined to birds. The Pediculi are flat, translucent, and short-legged. They seem to belong to almost every known animal, and, in some cases, more than one species are found on the same animal. With the Nirmi, whatever may be the natural prejudice entertained against them, under the microscope a world of beauty and variety is revealed in their forms and structure.

TABLE OF FAMILIES OF DIPTERA.

SECTION I.—PROBOSCIDEA.

Mouth having a more or less fleshy trunk or proboscis, bilobed at the tip.

Legs set close together on the under side of the thorax.

SUBSECTION I.—NEMOCERA.

Antennæ long, of six to ten distinct joints; hairy.

Wings; anal areolet open at the hind margin.

Figure usually slender, hump-backed, long-legged.

Families:

  a. Ocelli two or three.
    1. Discal areolet complete.
      Rhyphidæ.
    2. Discal areolet wanting.
      Mycetophilidæ (1st segment of thorax inconspicuous).
      Bibionidæ (1st segment of thorax conspicuous).

  b. Ocelli none.
    1. Costal vein ending near tip of wing.
      Simulidæ—*Sandflies.* (Tibiæ and 1st joint of tarsus broad and compressed.)
      Chironomidæ—*Midges.* (Tibiæ and 1st joint of tarsus slender, sub-cylindrical.)

376   INSECTS.

  2. Costal vein attenuated round tip of wing.
   * Veins in last subdivisions not more than six.
    Cecidomyzidæ.
   ** Veins in last subdivision more than six.
    Culicidæ—*Gnats, Mosquitoes* (Hind-margin fringed with scales.)
    Heteroclitæ. (Hind-margin fringed with hairs.)
    Phlebotomidæ. (Wings ovate, or lanceolate; deflexed or divaricated.)
 c. Middle segment of thorax with an angular, transverse seam.
  Tipulidæ—*Daddy Longlegs*, &c.

SUBSECTION II.—BRACHYCERA.

Antennæ of three distinct joints, from the third of which springs a bristle; or from three to ten joints, which, after the third, are fused together, or suddenly become abruptly slender.

Wings; anal areolet closed.

Figure usually stouter than in Nemocera.

 Families:

a. Cubital vein forked;† anal areolet tapering nearly or quite to hind-margin.
   1. Costal vein ending with tip of wing.
    Stratiomidæ. (*Soldier-flies.*) Antennæ five to ten joints, conical at tip, or terminating in long style or short bristle.
    Scenopinidæ. Antennæ three-jointed, spindle-shaped; third joint large and long.

---

 † In Empis the cubital vein is sometimes forked, sometimes simple; the anal areolet not reaching near hind margin.

2. Costal vein attenuated to hind margin.
* Alulæ very large.
Tabanidæ. (*Horseflies.*) Antennæ six, eight, or ten joints; joints unequal.
** Alulæ very small.
Xylophagidæ. Proboscis withdrawn; antennæ ten joints; joints equal after second, and cylindrical, or tapering.
Asilidæ. Foot-pads very large,† forehead hollowed. Antennæ three or five joints, unequal.
Leptidæ. Empodium dilated like foot-pads; forehead convex; antennæ generally three joints, third largest, transverse or conical, or bearing a bristle.
Bombylidæ. (*Beeflies.*) Empodium slender, pointed; forehead convex; antennæ three or four joints, unequal.
b. Cubital vein simple.
1. Axillary lobe of wing obsolete.
* Tip of wing rounded.
Empidæ. Brachial and anal areolets, some of them reaching to one-third of wing; abdomen seven or eight segments.
Dolichopidæ. Brachial and anal areolets minute or indistinct; abdomen five segments, or in ♂ 6, rarely seven.
** Tip of wing pointed.
Lonchopteridæ.
2. Brachial veins each accompanied by spurious veins.
Syrphidæ (*Droneflies*, &c.)

---

† Except in one genus, Leptogaster, to be recognised by the absence of foot-pads.

3. Brachial veins without spurious veins.
> Platypezidæ. Antennæ, third joint globose, conical or tapering, tipped by a bristle.
> Conopidæ. Antennæ long, no bristle; three first joints large, remainder small, pointed.
> Pipunculidæ. Antennæ bearing a bristle on the upper edge of third joint; eyes reaching to the mouth.
> Muscidæ. (*Housefly, &c.*) Antennæ bearing a bristle on the upper end of third joint; eyes bounded by the cheeks.

c. Halteres covered by alulæ, and head very minute. Acroceridæ.

d. Proboscis obsolete.
> Œstridæ. (*Gadflies.*)

SUBSECTION III.—HYPOCERA.

Antennæ very short, close to mouth, three jointed globose, bearing a bristle.

Veins of wings few, not branching.
> Family:
> Phora.

SECTION III.—EPROBOSCIDEA.

Mouth having a tubular proboscis.
Legs set on either side of a broad breastplate.
Antennæ buried in furrows near the mouth.
> Families:
> 1. Head large, sunk in the thorax.
>   Hippoboscidiæ—*Forest-fly, Sheep-tick, &c.*
> 2. Head small, thrown backwards, and held upside-down.
>   Nycteribiidæ.

The above characters are taken from Mr. Walker's work on Diptera.

TABLE OF FAMILIES OF DIPTERA. 379

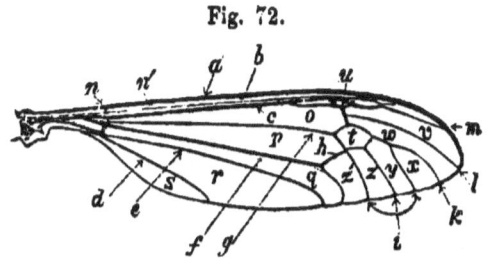

Wing of *Tipula*.

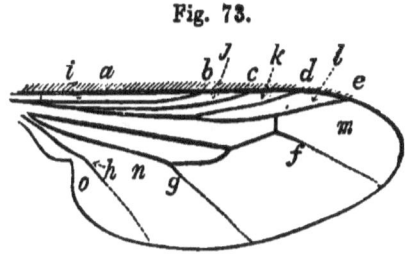

Wing of *Hippobosca*.

Nemo-cera.	Nervures.	Brachy-cera.	Nemo-cera.	Cells or Areolets.	Brachy-cera.
a —	Costal	— a		Mediastinal	— j
b —	Mediastinal	— b	n —	Sub-costal	— i
c —	Sub-costal	— c	o —	Præ-brachial	
d —	Axillary		p —	Post-brachial	
e —	Anal	— h	q —	Anal	— n
f —	Post-brachial		r —	Axillary	— o
g —	Præ-brachial		s —	Sub-axillary	
h —	Post-brach. *veinlet*		t —	Discoidal	
i i —	Externo-medial	— fg	u —	Radial	— k
k —	Sub-apical		v —	Cubital	— l
l —	Cubital	— e	w —	Sub-apical	— m
m —	Radial	— d	x y z z—	Externo-medial	
n —	Cross veinlet of antejugal axis				

GLOSSARY.

Abdomen, the last section of the body.
Aculeate, needle-like; or, furnished with a needle-like ovipositor.
Alulæ, or "little wings," a small membranous appendage at the base of each wing in Diptera.
Areolet, a cell, or enclosed space formed by the nervures of a wing.
Base, the part nearest the trunk of body.
Brachelytrous, having short elytra, or wing-cases.
Capitate, headed, or knobbed, fig. 2, p. 65.
Clavate, club-shaped.
Costal, belonging to the costa, or front edge of the wing, see figs., pp. 254, 379.
Coxa, the basal joint of the leg, fig. 9, *a*, p. 35.
Deflexed, bent down, or shelving, as the wings of a Lacefly.
Elytra, the horny wing-cases of Beetles, &c.
Entomophagous, insect-eating.
Femur, the first long joint of the leg, the thigh, p. 35, fig. 9, *c*.
Filiform, thread-like.
Fissate (antenna), cleft, see p. 65, fig. 31, 6.
Fossorial, digging.
Geniculated, bent like a knee.
Halteres, poisers, an appendage of flies, p. 54.
Hemelytra, the partially thickened fore-wings in Heteroptera.
Hind-margin of wing, the edge furthest from the body.
Imago, the insect in its final, or perfected state.
Inner margin of wing, that opposite the costa.
Labial palpi, feelers of the labium, p. 30, fig. 5 (Beetle); p. 225, figs. 58, 59, *c* (Bee).
Labium, the under lip, or tongue, p. 30; diagram, p. 29; p. 30, fig. 5 (Beetle); p. 32, fig. 6 (Bee).
Labrum, the upper lip, p. 30; diagram, p. 29.
Lamellate (antenna), leaf-like, p. 65, fig. 31, 5.
Larva, the insect in the first stage (*e.g.*, in Butterflies the Caterpillar), see Ch. IV.
Ligula, sometimes called tongue, p. 32, fig. 6, *g*; p. 225, fig. 59, *a*.
Malleoli, the halteres.
Mandibles, the upper pair of jaws, p. 29; p. 30, fig. 3.
Mandibulate, having jaws.
Maxillæ, the under pair of jaws, p. 29; p. 30, fig. 4 (Beetle); p. 225, fig. 58, *d* (Bee).
Maxillary palpi, feelers of maxillæ, p. 30, fig. 4 (Beetle); p. 225, figs. 58, 59, *e* (Bee).

GLOSSARY.

Mentum, chin, or base of tongue, p. 32; p. 225, fig. 58, fig. 59.
Metatarsus, the basal joint of the tarsus ; that next the tibia.
Metathorax, the third segment of the thorax, bearing the third pair of legs, and second pair of wings.
Mesothorax, the middle segment of the thorax, bearing the second pair of legs and first pair of wings.
Moniliform, like a string of beads.
Ocelli, simple eyes, p. 27.
Onychia, foot-pads.
Ovoid, egg-like.
Palpi, feelers.
Paraglossæ, filaments growing on the tongue in Hymenoptera, p. 225, fig. 58, *b* ; 59, *b*.
Pectinate, toothed, like a comb.
Perfoliate (antennæ), like leaves run through with a thread, p. 65, fig. 3.
Phytophagous, plant-eating.
Prosternum, front of the sternum.
Prothorax, first segment of the thorax bearing the first pair of legs.
Pseudo, false.
Pulvilli, foot-pads.
Pupa, the insect in its second stage ; in Butterflies the Chrysalis, see Ch. IV.
Rypophagous, filth-eating.
Scutellum, little shield, a triangular part of the mesothorax seen at the base of the elytra. Conspicuous in some Heteroptera.
Serrate, toothed like a saw.
Setaceous, bristle-like.
Shank, see Tibia.
Stemmata, simple eyes, p. 27.
Sternum, the breast-plate.
Stigma, a spot on the costa of the wing ; conspicuous in Neuroptera and Hymenoptera.
Sub-imago, see p. 135—7.
Tarsus, the foot, or last series of joints in the leg, generally ending in a pair of claws, p. 35, fig. 9, *e*, &c.
Tegmina, the roof-shaped thickened fore-wings in Orthoptera and Homoptera.
Tegulæ, a pair of large triangular scales fixed to the base of the fore-wings in Lepidoptera.
Thigh, see Femur.
Thorax, that section of the body which bears the legs and wings.
Tibia, the second long joint of the leg, or shank, p. 35, fig. 9, *d*.
Trifid, cleft in three.
Trochanter, the second joint of the leg, p. 35, fig. 9, *b* (sometimes consisting of two pieces), see p. 35.
Trophi, organs of the mouth.
Ungues, foot-claws.
♂, male.
♀, female.
☿, neuter.

INDEX TO FAMILIES, GENERA, &c.

ACALYPTA, 322
Acalypteræ, 366
Acanthia, 319
Acanthiidæ, 319
Acherontia (Death's-head Moth), 269
Acheta, 116
Achetidæ, 116
Acilius, 69
Aculeata, 187
Adephaga, 66
Æshna, 128
Agramma, 322
Agrion, 129
Aleyrodes, 308
Allantus, 158
Alucitina, 275
Ammophila, 206
Anax, 128
Andrena, 228
Andrenidæ, 225
Andrenoides, 235
Aneurus, 322
Anobium, 84
Anthidium, 240
Anthocharis, 260
Anthomyia, 366
Anthophila, 187, 221
Anthophora, 242
Anthrocera (Burnet Moth), 270
Apathus, 244
Apatura (Purple Emperor), 262
Aphaniptera, 61, 329
Aphelocheirus, 319
Aphidæ, 302
Aphis, 303
Aphrophora, 296
Apidæ, 231
Apis, 245
Aporus, 205

Aprosterni, 81
Apterygidæ, 110
Aradus, 321
Arctia (Tiger Moths), 270
Arge (Marbled-white Butterfly), 261
Argynnis, 263
Asilus, 356
Astemma, 323
Athalia, 158
Atherix, 357
Atrachelia, 88
Aurocorisa, 319

BERIS, 354
Bibionidæ, 345
Blaps, 88
Blatta, 114
Bombus, 243
Bombycidæ, 293
Bombycina, 270
Bombylidæ, 357
Bombylius, 357
Boreus, 141
Botys (Small Magpie), 274
Brachelytra, 75
Brachinus (Bombardier), 68
Brachycera, 340, 353
Bruchus, 90
Bucentes, 364
Byrrhus, 77

CÆLIOXYS, 234
Calandra, 90
Calepteryx, 129
Callimome, 183
Calypteræ, 354
Cantharis, 86
Capsidæ, 322
Capsus, 322

INDEX TO FAMILIES, GENERA, ETC. 383

Carabus, 67
Cassida, 94
Catocala (Red underwing), 272
Cecidomyia, 346
Cecidomyzidæ, 344
Cephalemyia, 370
Cerambyx, 92
Ceratina, 241
Cerceris, 212
Cercopidæ, 301
Ceropales, 205
Cerura (Puss Moth), 293
Cetonia, 79
Chærocampa (Elephant Hawks), 269
Chalcis, 182
Chelonidæ, 293
Chelostoma, 240
Chelymorpha, 307
Chlorops, 368
Chironomidæ, 346
Chironomus, 346
Chrysis, 184
Chrysopa, 139
Chrysophanus, 265
Cicada, 299
Cicindela, 66
Cimbex, 157
Cimicidæ, 320
Cixius, 301
Clavicornes, 77
Claviger, 76
Clytus, 93
Coccinella (Ladybird), 95
Coccus, 309
Cochlipodidæ, 293
Cœnonympha (small Heath Butterfly), 262
Coleoptera, 6C, 63
Colias, 260
Colletes, 225
Conopidæ, 363
Conops, 363
Cordylocerata, 77
Coreidæ, 323
Corixa, 317
Cossus (Goat Moth), 270
Crabro, 209
Crabronidæ, 209
Crioceris, 93
Cuculinæ, 235
Culicidæ, 347

Culex (Gnat), 347
Cydnus, 325
Cymus, 324
Cynips, 172, 177
Cynthia (Painted Lady), 262

DASYCHIRA (TUSSOCK), 270
Dasygastræ, 236
Dasypoda, 230
Dermestes, 74
Delphax, 301
Dimera, 302
Diodontus, 211
Diploptera, 213
Diptera, 61, 322
Dolichopidæ, 359
Dosytheus, 158
Drosophila, 368
Dyticus, 69

EGERIIDÆ, 292
Elater, 80
Empidæ, 358
Empis, 358
Endromidæ, 293
Entomophaga, 169
Epeolus, 234
Ephemeron, 135
Ephippiger, 121
Ephyridæ, 288
Eproboscidea, 372
Eristalis, 264
Erycinidæ, 264
Eucera, 241
Eumenes, 213
Eumenidæ, 212
Euplexoptera, 60, 109
Eurydema, 324
Evania, 178
Evaniidæ, 178

FORFICESILA, 110
Forficula, 110
Formicidæ, 190
Fossores, 203
Fulgoridæ, 301

GALLERIDÆ, 289
Gasterophilus, 371
Gastrodes, 323
Geodephaga, 66
Geometrina, 272

INDEX TO FAMILIES, GENERA, ETC.

Geotrupes, 78
Gerris, 320
Goërius, 75
Gonepteryx, 260
Gonoptera, 272
Gorytes, 208
Grapta, 263
Gryllidæ, 120
Gryllotalpa, 116
Gyrinus, 70

HALICTUS, 227
Haltica, 94
Hedychrum, 186
Helophilus, 360
Hemerobius, 139
Hemiptera, 297
Hepialus (Swifts), 271
Heriades, 241
Hesperidæ, 265
Heterocera, 268
Heteroclitæ, 350
Heterogyna, 189
Heteromera, 86
Heteroptera, 61, 315
Hilara, 359
Hipparchia (Meadow Brown Butterfly), 262
Hippobosca, 372
Hister, 77
Homoptera, 60, 296
Hydradephaga, 68
Hydrocorisa, 316
Hydrometra, 320
Hydrometridæ, 319
Hydrophilus, 71
Hydropsychides, 150
Hydroptilides, 150
Hydrous, 71
Hylotoma, 157
Hymenoptera, 61, 152
Hypena, 273
Hypera, 159
Hypocera, 348, 371

ICHNEUMON, 179
Ichneumonidæ, 178

LABIA, 110
Lamellicornes, 77
Lampyris (Glow-worm), 81
Larridæ, 206

Lepidoptera, 255
Leptidæ, 356
Leptis, 357
Leptocerides, 151
Libellula, 128
Limenitis (White Admiral), 262
Liparidæ, 293
Lithosidæ, 293
Locustidæ, 121
Longicornes, 92
Lucanus (Stag Beetle), 77
Lycænidæ, 264
Lygæidæ, 323
Lygæus, 323

MACROSTERNI, 81
Malachius, 83
Megachile, 239
Melecta, 234
Melitæa, 263
Melithneptus, 362
Mellinus, 208
Meloe, 87
Melolontha (Cockchafer), 79
Melophagus, 372
Membracis, 302
Mimesa, 211
Miscophus, 207
Molorchus, 93
Monomera, 309
Musca, 363
Muscidæ, 363
Mutilla, 202
Mutillidæ, 190, 201
Mycetophilidæ, 344
Myrmecidæ, 190, 201

NAUCORIS, 318
Necrophaga, 72
Necrophorus, 73
Nematus, 158
Nemeobius, 264
Nemocera, 339, 343
Nepa, 318
Nepidæ, 318
Neuroptera, 60, 126
Nirmi, 373
Noctuina, 271
Nomada, 234
Notodontidæ, 293
Notonecta, 316
Notonectidæ, 316

INDEX TO FAMILIES, GENERA, ETC. 385

Nycteribia, 372
Nymphalidæ, 261
Nysson, 207
Nyssonidæ, 207

ODYNERUS, 313
Œdemera, 88
Œstridæ, 369
Œstrus, 369
Ophion, 180
Orgyia (Vapourer), 270
Ortalis, 367
Orthoptera, 60, 113
Osmia, 236
Oxybelus, 209

PANORPA, 140
Panurgus, 235
Papilionidæ, 260
Pediculi, 373
Pemphredon, 211
Pentamera, 66
Pentatoma, 324
Perla, 138
Philanthidæ, 211
Philanthus, 212
Philhydrida, 71
Phillophorus, 307
Phlebotomidæ, 350
Phora, 371
Phryganea, 146
Phryganeides, 151
Phyllopertha, 79
Phytophaga, 93, 156
Pieris, (White Butterflies), 261
Platypterigidæ, 293
Ploa, 317
Plusia, 272
Podops, 324
Polyommatus (Blue Butterflies), 264
Pompilidæ, 205
Pompilus, 205
Poneridæ, 190, 201
Prædones, 188
Priocerata, 86
Proboscidea, 330
Proctotrupes, 184
Prosopis, 225
Pselaphus, 104
Pseudo-tetramera, 88
Pseudo-trimera, 95
Psocus, 142

Psyche, 271
Psychidæ, 293
Psychoda, 350
Psychomides, 150
Psylla, 303
Psyllidæ, 302
Pterophorina, 275
Pulex (Flea), 329
Pyralidina, 273
Pyrochroa, 86

RAPHIDIA, 141
Ranatra, 319
Reduviidæ, 320
Reduvius, 320
Rhamphomyia, 359
Rhingia, 362
Rhopalocera, 359
Rhopalus, 324
Rhyncophora, 89
Rhyphidæ, 352
Ripiphorus, 87
Rumia, 272
Ryacophilides, 150
Rypophaga, 71

SALPINGIDÆ, 88
Sapyga, 204
Sapygidæ, 204
Sarcophaga, 365
Sargus, 354
Saropoda, 242
Saturnia (Emperor Moth), 271
Saturnidæ, 293
Scatophaga, 366
Scenopinidæ, 358
Scoliidæ, 204
Scolytus, 91
Scopelosoma (Satellite), 282
Scopulipedes, 241
Scutelleridæ, 324
Selandria, 165
Sepsis, 369
Sericostomides, 150
Serrifera (Saw-bearers), 156
Sesia (Clear-winged Moths) 270
Sialis, 140
Sigara, 317
Silona, 159
Silpha, 73
Simulidæ, 346
Sirex, 166

C C

INDEX TO FAMILIES, GENERA, ETC.

Sociales, 243
Sphesia, (Clear-winged Moths) 270
Sphecodes, 227
Sphegidæ, 206
Sphingina, 268
Sphinx, 269
Spiculifera, 170
Spilosoma (Ermine Moth), 271
Stelis, 234
Stomoxys, 364
Strangalia, 93
Stratiomidæ, 354
Stratiomys, 354
Strepsiptera, 101
Stylops, 101
Syrphidæ, 360
Syrphus, 361

TABANIDÆ, 355
Tabanus, 355
Tachines, 364
Tachytes, 207
Telephorus (Soldiers and Sailors), 81
Tenthredinidæ, 156
Tenthredo, 156
Tephrites, 367
Terebellifera (Borers), 166
Terebrantia, 155
Tetramera, 88
Thecla, 264
Thrips, 123
Thymele, 265
Thysanoptera, 60, 123

Timarchia (Bloody-nosed Beetles), 94
Tineina, 275
Tingidæ, 321
Tiphia, 204
Tipula, 350
Tipulidæ, 350
Tortricina, 274
Tortrix, 294
Trachelia, 86
Trichiosoma, 157
Trichoptera, 60, 146
Trimera, 95, 299
Trochilium, 292
Trypeta, 367
Trypoxylon, 209
Tubulifera, 184

UROCERIDÆ (Wood-borers), 166

VANESSA, 262
Velia, 320
Vespa, 215
Vespidæ, 214
Volucella, 361

XYLOPHAGIDÆ, 355
Xylophagus, 355

ZANTHOSETIA, 274
Zeuzera, 270
Zeuzeridæ, 292
Zygænidæ, 292

GENERAL INDEX.

ADMIRAL butterflies, 262
Aleppo galls, 177
American blight, 305
Ants, 188
  as architects, 197
  as cattle-owners, 194
  as friends, 191
  as nurses and parents, 192
  as slave-owners, 195
  as soldiers, 195
  care for their dead, 192
  do not store grain, 200
  used in divination, 8
  in Eastern mythology, 8
  (tradition concerning) in Ceylon, 8
  (social), 190
  (solitary), 201
  (larvæ of), 56
Ant baths, 198
Ants' nests, acid in, 198
  ,, beetles in, 76—199
Ant vinegar, 199
Antennæ of insects, 28
Aphis, 303
  wings of, 52
  in Polar seas, 306
Asparagus beetle, 93
  larva of, 99
BACON beetle, hair of, 74
Bark-lice, 309
Bark-mining beetle, 90
Bath-white butterfly, 261
Bees (and *see* Hive Bees),
  table of, 253
  cuckoo, 222, 231, 244
  hairy-bellied, 236
  hairy-legged, 241
  hive, 245
  humble, 243
  temper of, 244
  leaf-cutter, 239

Bees, long-tongued, 231
  mason, 236
  parasitic, 222, 231, 244
  short-tongued, 224
  eyes, 26, 357
  legs, 222, 248
  mouth, 32
  wings, 49
  ,, hooks on, 50, 203
  larvæ, 56
  pupæ, 57
  nests, 236
  food of, 249
  products of, 248
  enemies of, 249
  stingless, in Mexico, 247, *n.*
  how to find specimens of queen, 248
  in Egyptian hieroglyphs, 4
  in the East a symbol of fecundity, 5
  on Grecian and Ephesian coins, 6
  omens of future eloquence, 7
  used as a badge by ancient kings of France, 4
  superstitions concerning, 7
  in heraldry, 8
Beebread, 248
Beefly, 357
Beetles (and *see* Water Beetles)
  table of, 103
  bacon, 74
  bloody-nosed, 94
  bombardier, 68
  burying, 72
  death-feigning, 77
  dung, 78
  musk, 92
  oil, 87
  pill, 77
  scavenger, 72

GENERAL INDEX.

Beetles, sexton, 72
   water, 68, 71
   wood-boring, 84, 90
   legs, 35, 36
   mouth, 29
   wings, 41
   larvæ, 56
   pupæ, 57
   in ants' nests, 76
Birdlice, 373
Blackbeetles, 114
Black-veined white butterfly, 261
Bloody-nosed beetle, 94
Blowfly, 365
Bluebottle, 365
Blue butterflies, 264
Bombardier beetle, 68
Book-louse, 142
Book-worm, 84
Botfly, 369
Breathing apparatus of insects, 21
Breezefly, 369
Brimstone butterfly, 260
   moth, 272
Brown wood butterflies, 262
Brown (meadow) butterflies, 262
Buff-tip moth, 270
Bug (bed), 230
   eaten by cockroaches, 115
   (plant), 315
Burnet moths, 270
Burying beetles, 72
Butterflies,
   (table of), 290
   Bath white, 261
   black-veined white, 261
   blue, 264
   brimstone, 260
   brown wood, 262
   clouded yellow, 260
   copper, 265
   fritillary, 263
   marbled white, 261
   meadow brown, 262
   orange tip, 260
   painted lady, 262
   peacock, 262
   purple emperor, 262
   red admiral, 263
   small heath, 262
   sulphur, 260
   swallow-tail, 260

Butterflies, tortoiseshell, 263
   white, 260
   white admiral, 262
Butterflies' antennæ, 259
   eyes, 26
   mouth, 33
   scales, 51
   wings, 50, 259
   changes, 55
   larvæ, 284
   pupæ, 284
   swarms of, 266
Butterflies and moths, to distinguish, 257—259
Butterfly, emblem of the soul, 10
CABBAGE butterfly (white), 260
Caddis-fly, 146
   ,, wings, 46
   worms, 146
Camberwell beauty, 267
Cardinal beetles, 86
"Carpets" 294
Case-bearing larvæ of beetle, 100
   ,, caddis, 147
   ,, moths, 279
   ,, sawflies, 165
Chafers, 77
Changes of insects, 22, 55
Characters of insects, 14
Cheese-hopper, 368
Chigoe, 331
China-mark moth, 294
Cholera-fly, 332
Churchyard beetle, 88
Chrysalis the husk of a butterfly, 23
Cicada, 299
   saws of, 301
   sings till it bursts, 300
   victor in musical contest, 9
Circulating system of insects, 21
Clear-winged moths, 270
Clegs, 355
Clothes-moths, 275
   larvæ of, 279
   ,, morals of, 280
Clouded yellow butterfly, 260, 266
Cochineal insect, 310
Cockchafer, 79
   larvæ of, 98
   wings of, 41
   Swedish superstitions concerning, 13

GENERAL INDEX.

Cockroach, 114
  egg-case of, 115
  larvæ of, 56
Coleoptera (*see* Beetles), 63
  table of, 103
  antennæ, 65
  legs, 35
  wings, 41
  larvæ, 56, 98
  pupæ, 57
Concealment, self, of larvæ of asparagus beetle, 99
  ,, bloody-nosed beetle, 100
  ,, caddis, 146, 147
  ,, clothes-moths, 279
  ,, tortoise beetles, 10
Convolvulus hawk-moth, 269
Copper butterfly, 265
Corn weevil, 90
Courtilière, 118
Cranefly, 350
Creaking beetle, 94
Cricket, 114, 116
  musical instrument of, 43
  song sharp before rain, 119
Cuckoo bees, 224, 244
Cuckoo spit, 296
Currant moth, 273
DADDY-LONGLEGS, 350
Dagger moth, 294
Death's-head moth, 269
  feeds on bees, 249
Destruction of trees by wood-boring beetles, 91
  of crops by turnip-fly,
Devil's coach-horse, 75
  larvæ, 99
Diamond beetle, 89
Dragon-flies, 127
  eyes, 27
  flight, 128, 130
  metamorphosis, 131
  wings, 45
  larva, mouth of, 131
Drinker moth, 270
Dronefly, 360
  early life of, 46
Dung-beetle, 77
Dungfly, 366
EARTH-MEASURING caterpillar, 272
Earwigs, 109
  eyes, 27

Earwigs, larvæ, 56
  pupæ, 58
  wings, 43
  destroy bees, 249
  good mothers, 111
Eggs of lacefly, 139
  of whirligig beetle, 71
  of insects grow, 163
Elephant-hawk-moth, 269
Emperor-moth, 271
  larva, 276
Enemies of aphides, 303
  of bees, 249
Ermine moth, 271
Eyes of insects, 25
  bipartite, 70, 308
  of larvæ, 27
  of whirligig beetle, 70
Feet of insects, 38
Figure-of-8 moth, 293
Fireflies, 301
  superstition concerning, 12
Fleas, 53, 329
  legend concerning, 331
Flies, 332
  mouth of, 34, 336
  wings of, 46, 54
  larvæ, 337
  pupæ, 57, 353
  worship of, 1
Flight of insects in connexion with their vision, 26
  butterflies, 266
  dragon-flies, 128
  gnats, 349
  syrphidæ, 361
Flower-flies, 366
"Footman," 293
Forest-fly, 372
Formic acid, 198
Fritillary, 263
Froghopper, 296
GADFLY, 355, 369
Gallflies, 172
  gnats, 344
Galls, 173
Galls of Hymenoptera harbouring other insects, 176
  formed by Dipterous flies, 341—344
Gall-like excrescences formed by sawflies, 163

GENERAL INDEX.

Glowworm, 81
  luminous after death, 82
  larva luminous, 82
Gnats, 347
Goat-moth, 270
  larva, 280, 282
Grasshoppers, 114, 120
  legs, 37
  wings, 43
  vision, 27
  larvæ, 56
  pupæ, 58
  among the Greeks, 9
Grass-moths, 273
Green hair-streak, 264
Greenbottle-fly, 365
Ground-wasps, 219
HAIR of bacon beetle, 74
Hair-streak butterfly, 264
Hawk-moths, 269
Hessian-fly, 345
Hive-bee, 245
  legs of, 221, 248
  products of, 248
  enemies of, 249
  revengeful, 247
  in Peru ceased to lay up stores, 246
  queen, how to find specimens, 248
Honey, 248
Honey-dew, 306
Hop-aphis destroyed by ladybird, 97
Hop-dog, 276
Horsefly, 355
Housefly, 364
Humble-bees, 243
  jaws of, 232
  nests of, 243
  temper of, 244
Humming-bird moth, 270
ICHNEUMONS, 178
JAPAN Varnish, 311
Jigger, 331
June-bug, 79
KENTISH Glory, 293
Kurdish legend, 331
LAC, 311
Lacefly, 139
  eggs, 139
  larva, 303
Lacquer varnish, 311
Ladybird, 95

Ladybird (larvæ of), useful, 97, 304
  visitations of, 95
Lake, 311
Larvæ of beetles, 98
  (case-bearing), 100, 147, 165, 279
  (eyes of), 27
  of Hymenoptera, 153
  of laceflies, 140
  leaf-mining, 100
  of Lepidoptera, 265, 276
  of moths and butterflies, 283
  of moths in beehives, 250, 282
  of sawflies, 158, 164
  (wood-boring), 280
Leaf-cutter bees, 239
Leaf insects, 113
Leaf-mining larvæ, various, 281
  beetles, 100
  flies, 281
  moths, 277—280
Lead bored by beetles, 84
  by sirex, 168
Legs of Aculeate Hymenoptera, 187
  of bees, 222
  of butterflies and moths, 260
  of insects, 34
  of insects compared with those of vertebrate animals, 19
  of sawfly, 157
  of wasp, 187
Leopard moth, 270
  abounding, 267
  larvæ of, 280
Lice, 373
Locusts, 114, 122
  (foreign) in England, 122
Long-horned bee, 241
  beetles, 92
Long-tongued bees, 231
Loopers, 286
MAGPIE (or currant) moth, 273
  (small), 274
Mason bee, 236
  and ruby-tail, the biter bit, 186
May-bug, 79
May-fly, 135
Meadow-brown butterfly, 262
Mealworm, 88
Mealy bug, 309
Metamorphoses of insects, 22, 55
Midges, 346
Mole cricket, 117

Mole cricket, leg of, 37
Moths, 268, 357
  antennæ, 259, 269, &c.
  legs, 260
  mouth, 34
  wings, 50, 256
  position of, 259
  larvæ of, 276
  larvæ of in beehives, 250, 282
  pupæ of, 57
  in chrysalis three and six years, 267
Mosquitoes, 346
Mouths of insects, 29
Musk beetle, 92
NERVOUS System of Insects, 21
OAK Beauty, 294
  eggar, 293
Oil beetle, 87
Orange-tip Butterfly (Anthocharis), 260
Orders, table of, 60
Oxfly, 355, 370
PAINTED Lady, 262
Parasitic bees, 223, 244
  beetles, 87
  flies, 341
  Hymenoptera, 169
  Stylops, 102
Parasitism, 169
Peacock butterfly, 262
Pearls (moths), 294
Pea weevil, 90
Pill beetles, 77
Pitfalls dug by larvæ of devil's coach-horse, 99
  of tiger beetle, 99
  of fly, 357
Plume-moths, 275
Praying mantis, 113
Printer beetle, 92
Privet hawk-moth, 269
Prominent moths, 293
Propolis, 249
Pugs, 294
Pupæ of butterflies and moths, 282
Purple emperor, 264
  hair streak, 262
Puss-moth, 293
RAT-TAILED larva, 47, 57, 341
Red admiral, 263
  underwing, 272

Rose-aphis, 296
  Chafer, 79
Rubytail, 184
SAILORS, 81
Sand-flies, 346
Sand-wasps, 303
Satellite, 294
  larva of, 282
Saw of cicada, 300
  of sawfly, 161
Sawfly, 156
  (eggs of) grow, 163
  (swarm of), 159
  mother guards her young, 166
Scale insects, 309
Scarabæus, sacred beetle of the Egyptians, 13, 78
Scarlet-in-grain, 311
Scavenger beetles, 72
  flies, 333
Scorpion-fly, 140
Sexton beetles, 72
Sheeptick, 372
Short-tongued bees, 224
Silkworm moths, 270
Skeleton of insects, 19
Skipjack, 80
Skippers, 266
Small heath butterfly, 262
  magpie moth, 274
Snakefly, 141
Snout moths, 273
Soldier-flies, 354
Soldiers and sailors, 81
Spanish-fly, 86
Sphinx moths, 269
  larvæ, 269, 285
  in chrysalis three years, 267
Spiracles of insects, 22
  of water beetle, 39
Springtail, 372
Stable-fly, 364
Stag beetle, 77
Strawberry plume moth, 275
Stonefly, 139
Structure (external) of insects, 24
Sulphur butterfly, 260
Swallowtail butterfly, 258
  moths, 272
Swarms of ants, 159
  aphides, 306
  beetles (Curculionidæ), 158

GENERAL INDEX.

Swarms of butterflies, 266
 earwigs, 109
 gnats, 348
 insects near Deal, 159
 ladybirds, 95
 locusts, 122
 mayfly, 137
 turnip-fly, 158
Swifts, 271
Syrphus larva, 304
TABLE of Coleoptera, 103
 Diptera, 375
 Hymenoptera, 251
 Neuroptera, 144
 orders, 60
 Trichoptera, 150
Thrips, 123
Tiger beetle, 66
 ferocity of, 67
 mouth of, 30
 traps dug by larvæ of, 99
 moths, 270
 ,, larvæ, 286
Tithonus and grasshopper, 9
 (figure of), 113
Tortoise beetle, 94
 larva, 99
Tortoise-like leaf-bearer, 307
Tortoiseshell butterfly, 263
Tree wasps, 219
Troutfly, 135
Tongues of bees, 233, &c.
Turnip-fly, 94, 158
Tussoch moths, 270
 larvæ, 276
Twenty-plume moth, 275
VAPOURER moth, 270
Violet beetle, 67
WAINSCOT moths, 294

Walking-stick insects, 113
"Wasp-bees," 234
"Wasp-beetle," 93
Wasps, 213
 solitary, 213
 social, 214
 nest of, 215, 219
 love for young, 215
 stinging of, 218
Water beetle, 68
 legs, 35
 larvæ, 56
 nest, 72
 spiracles, 39
 ferocity of, 69
Water boatmen, 316
 leg, 36
Water measurers, 319
Water scorpion, 318
 pincers, 38
Wax, 248
Weevils, 89
Whirligig beetle, 70
 oar, 36
White admiral, 262
 ant, 126
 butterflies, 260
Willow-fly, 139
Wings of insects, 40
 compared with birds' wings, 20
Wireworm, 80
Wood-borers (Hymenopterous), 16
 (Coleopterous), 84, 90
Wood wasps, 203
Woolly bear, 286
Worship of the fly (ancient), 1
 (Hottentot), 2
YELLOW Ophion, 180
Yellow Sally, 139

THE END.

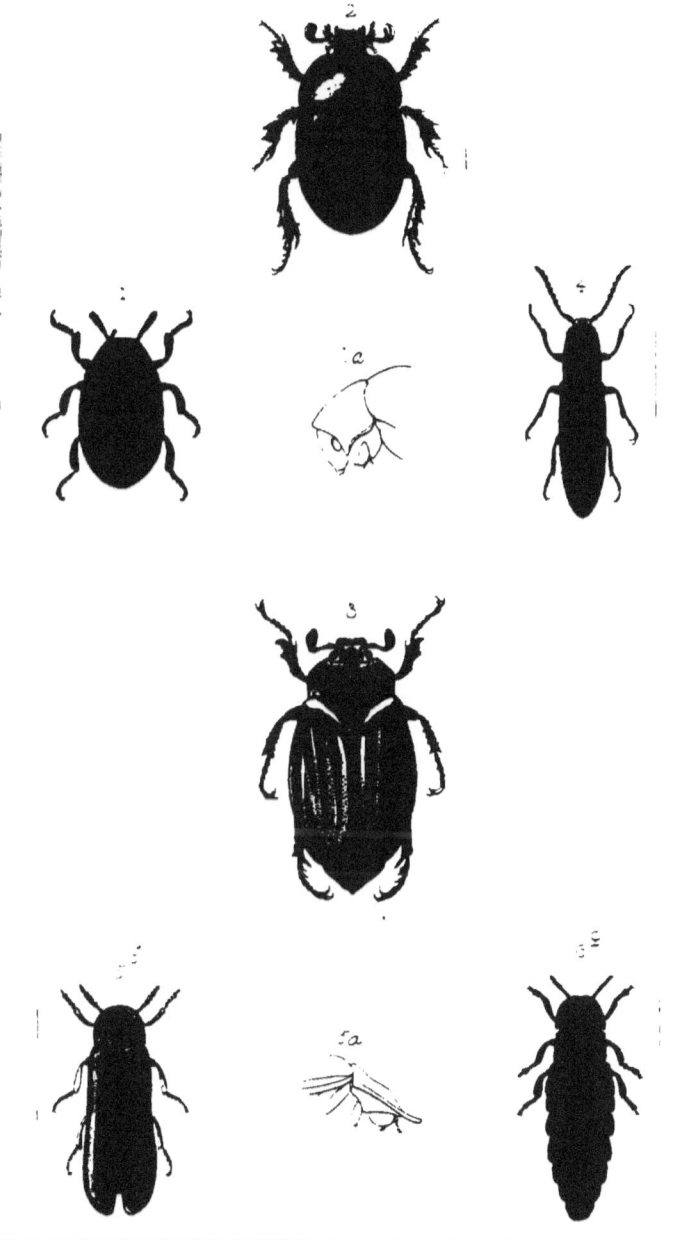

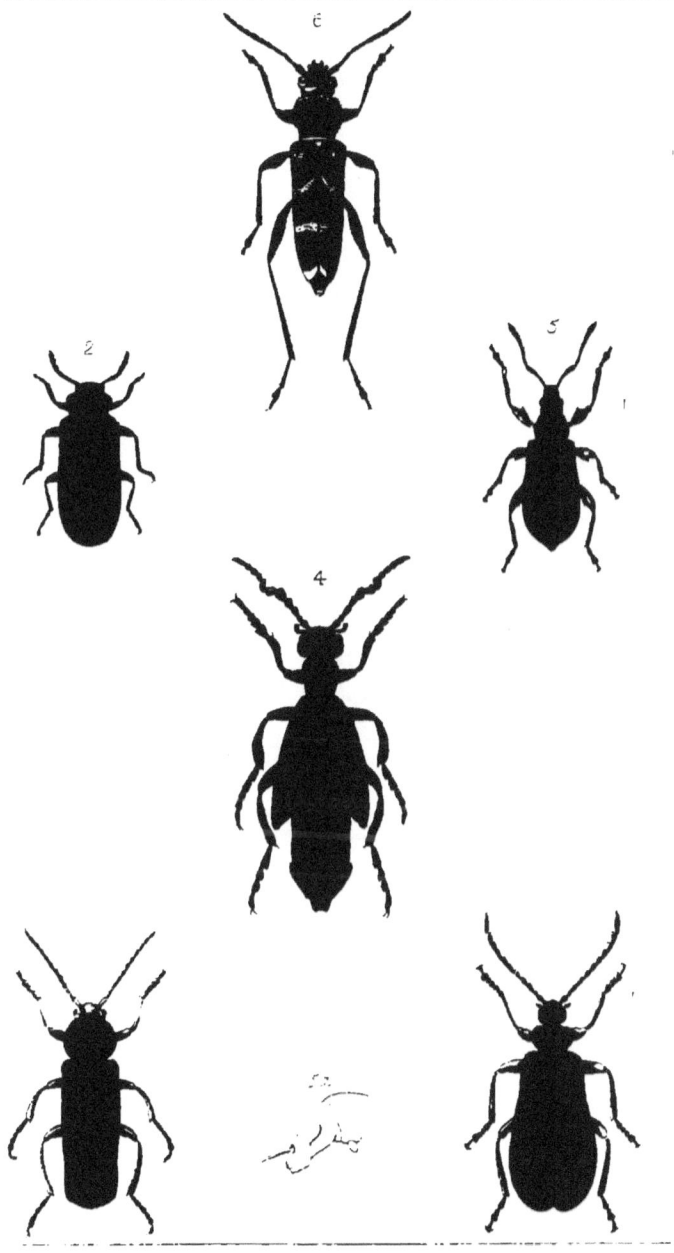

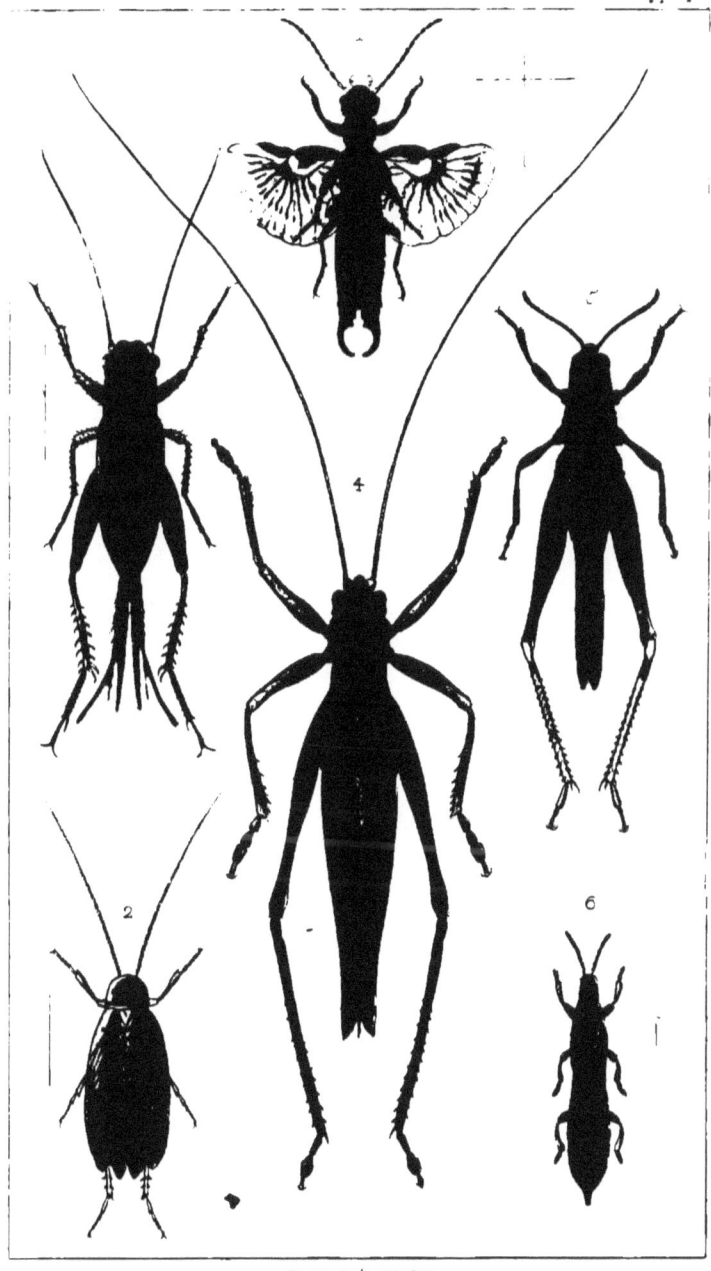

F.W.R. Del.et St 1870

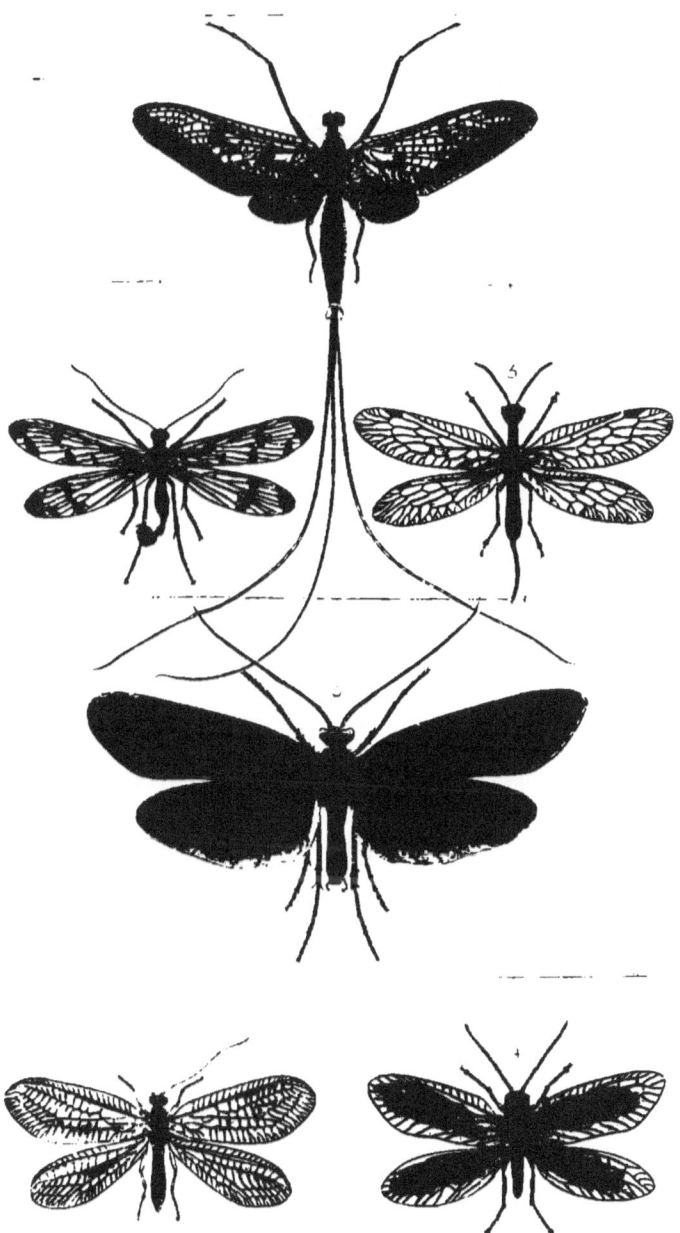

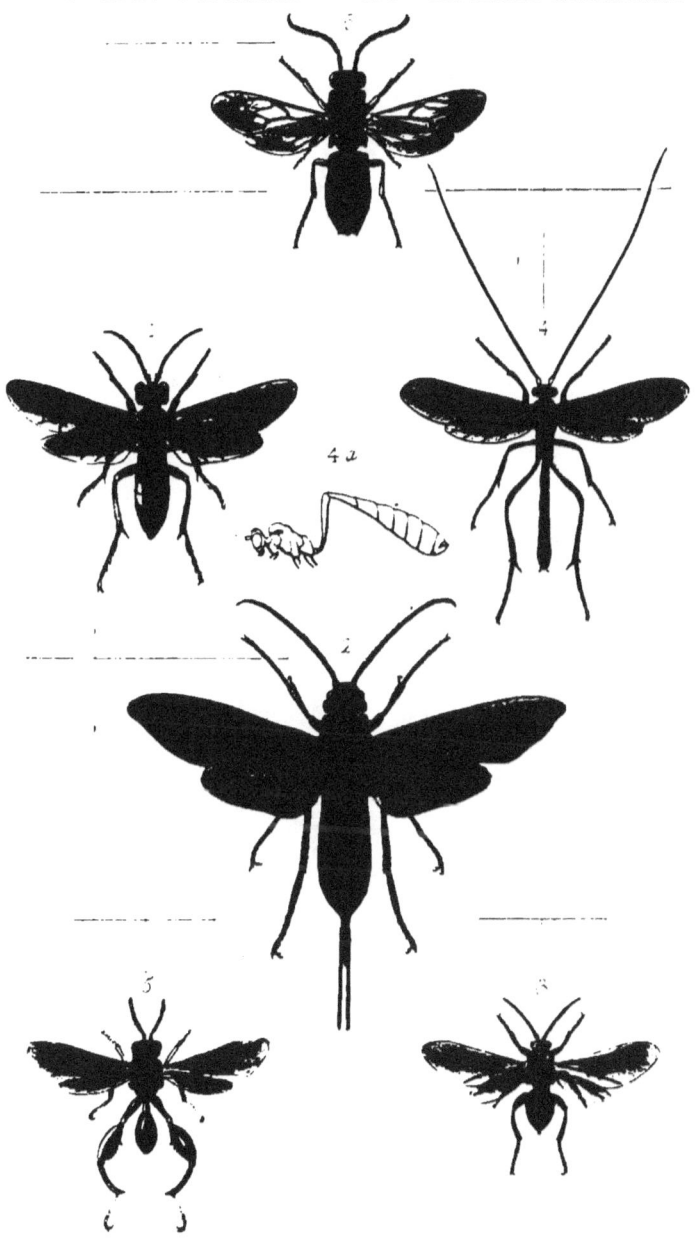

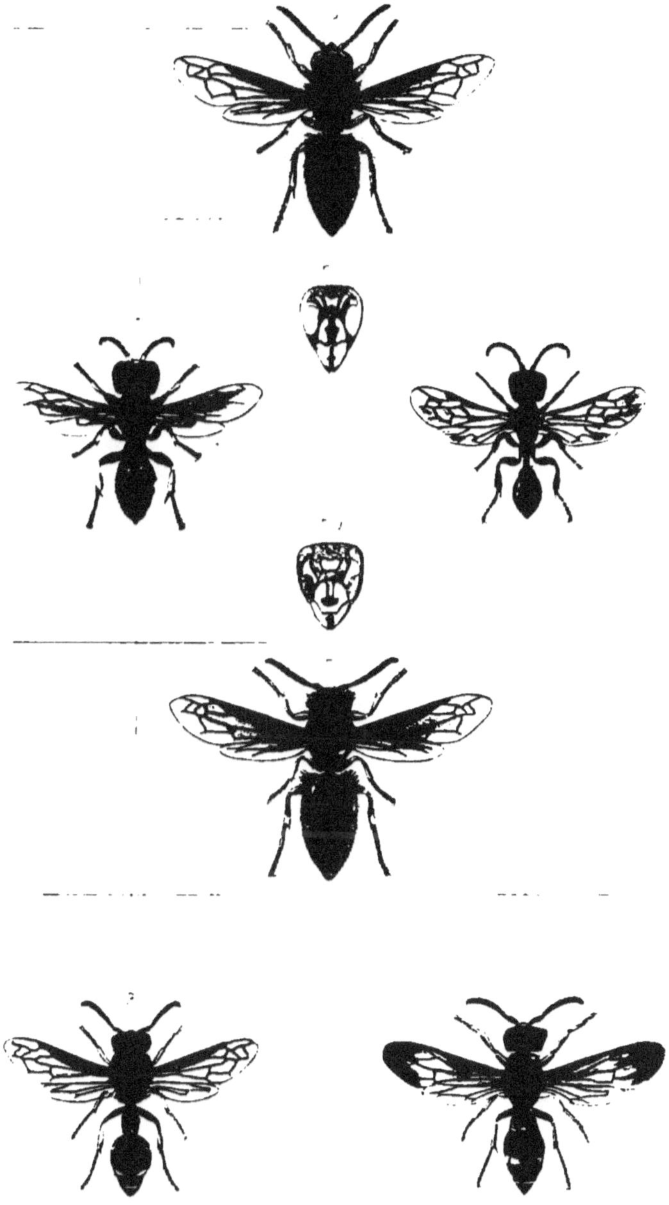

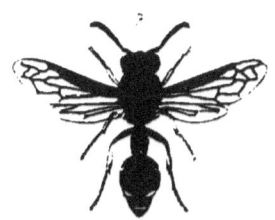

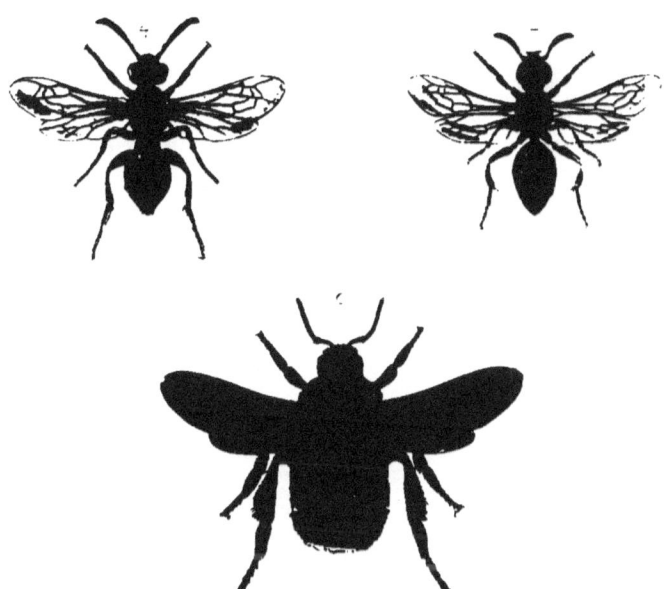

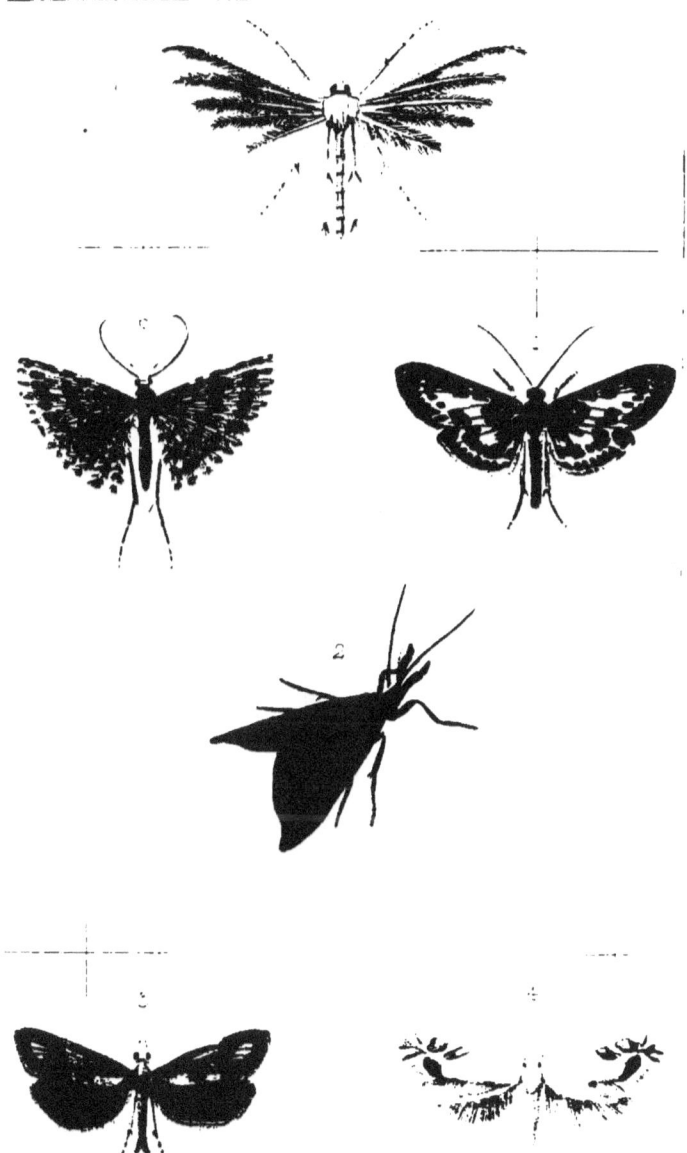

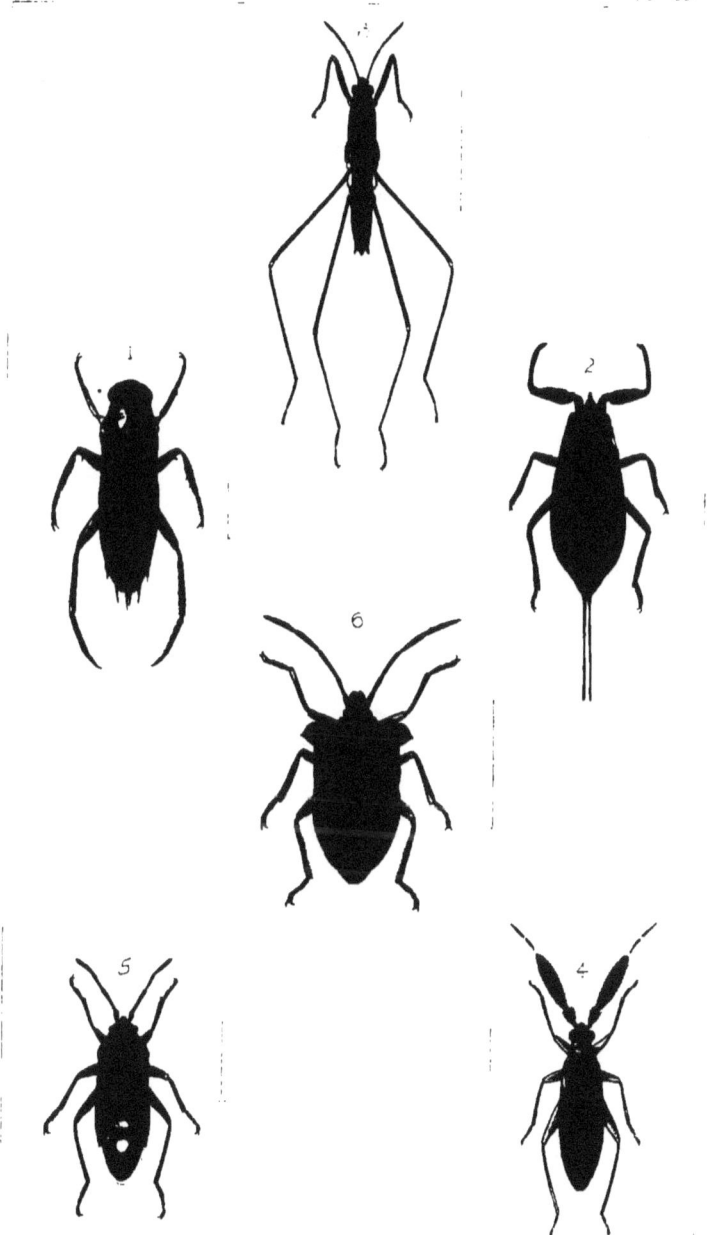

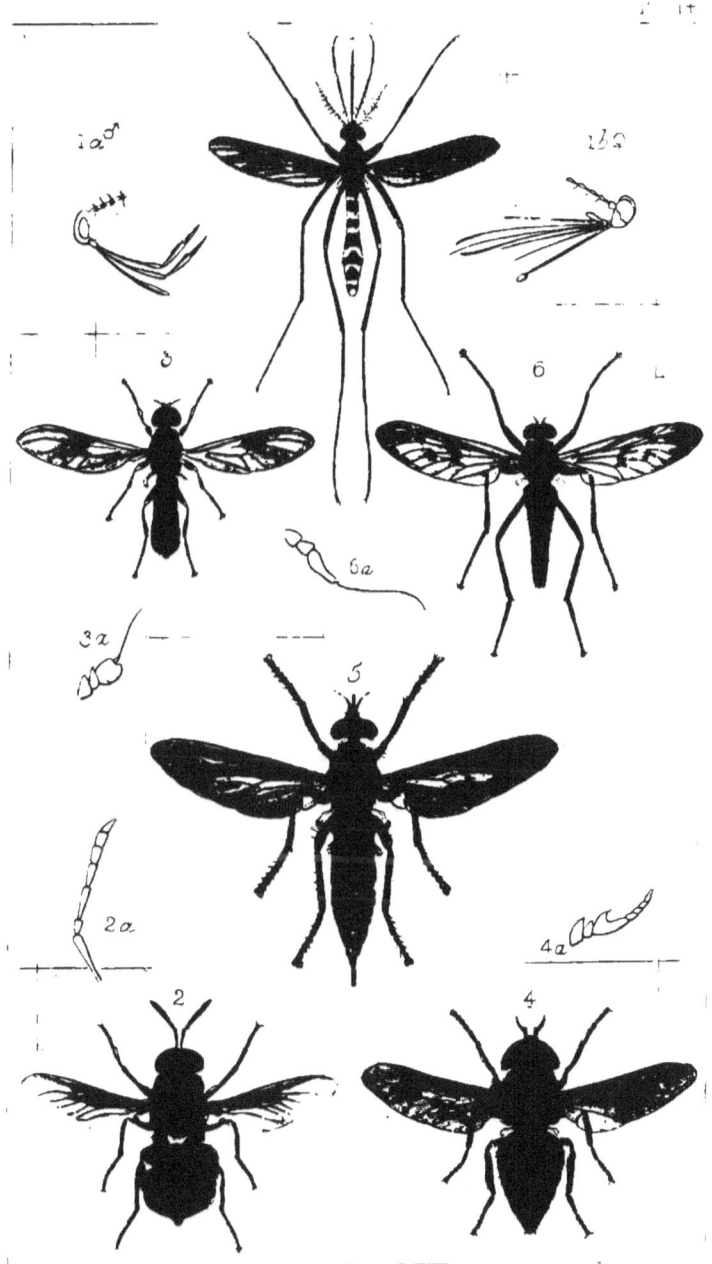

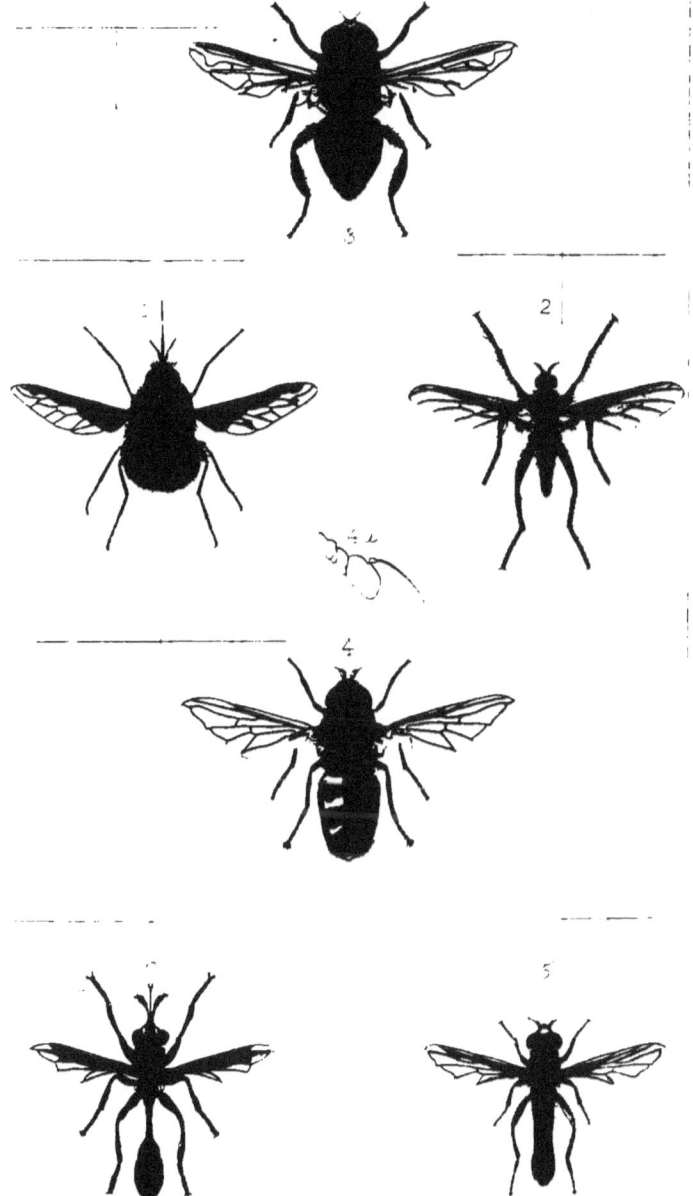

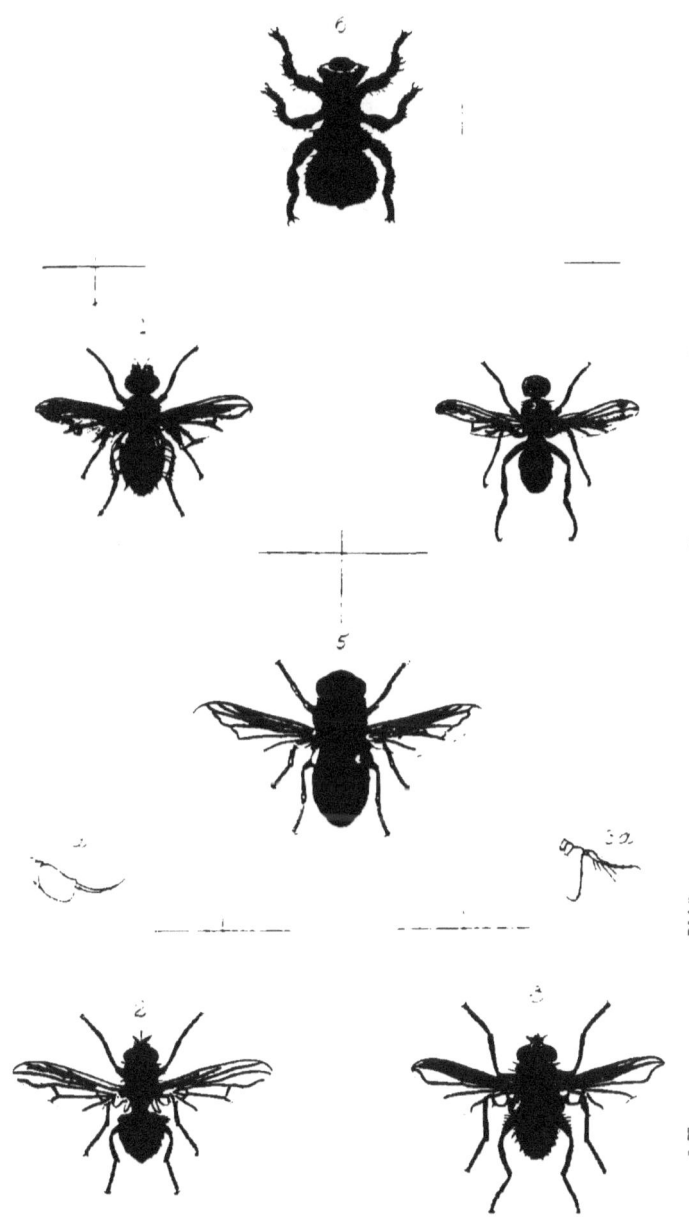

LIST OF WORKS

ON

BOTANY, ENTOMOLOGY, CONCHOLOGY,

TRAVELS, TOPOGRAPHY,

ANTIQUITY, AND MISCELLANEOUS

LITERATURE AND SCIENCE.

PUBLISHED BY

L. REEVE AND CO.,

5, HENRIETTA STREET, COVENT GARDEN, W.C.

NEW SERIES OF POPULAR NATURAL HISTORY FOR BEGINNERS AND AMATEURS.

British Insects; a Familiar Description of the
Form, Structure, Habits, and Transformations of Insects. By
E. F. STAVELEY. Crown 8vo, 16 Coloured Plates, and numerous
Wood Engravings, 14s.

British Butterflies and Moths; an Introduction to
the Study of our Native LEPIDOPTERA. By H. T. STAINTON.
Crown 8vo, 16 Coloured Plates, and Wood Engravings, 10s. 6d.

British Beetles; an Introduction to the Study of
our indigenous COLEOPTERA. By E. C. RYE. Crown 8vo, 16
Coloured Plates, and 11 Wood Engravings, 10s. 6d.

British Bees; an Introduction to the Study of the
Natural History and Economy of the Bees indigenous to the
British Isles. By W. E. SHUCKARD. Crown 8vo, 16 Coloured
Plates, and Woodcuts, 10s. 6d.

British Spiders; an Introduction to the Study of
the ARANEIDÆ found in Great Britain and Ireland. By E. F.
STAVELEY. Crown 8vo, 16 Coloured Plates, and 44 Wood
Engravings, 10s. 6d.

British Grasses; an Introduction to the Study of
the Grasses found in the British Isles. By M. PLUES. Crown
8vo, 16 Coloured Plates, and 100 Wood Engravings, 10s. 6d.

British Ferns; an Introduction to the Study of the
FERNS, LYCOPODS, and EQUISETA indigenous to the British
Isles. With Chapters on the Structure, Propagation, Cultivation,
Diseases, Uses, Preservation, and Distribution of Ferns. By
M. PLUES. Crown 8vo, 16 Coloured Plates, and 55 Wood
Engravings, 10s. 6d.

British Seaweeds; an Introduction to the Study of the Marine ALGÆ of Great Britain, Ireland, and the Channel Islands. By S. O. GRAY. Crown 8vo, 16 Coloured Plates, 10s. 6d.

BOTANY.

The Natural History of Plants. By H. BAILLON, President of the Linnæan Society of Paris, Professor of Medical Natural History and Director of the Botanical Garden of the Faculty of Medicine of Paris. Super-royal 8vo. Vols I. to VII., with 3200 Wood Engravings, 25s. each.

Handbook of the British Flora; a Description of the Flowering Plants and Ferns indigenous to, or naturalized in, the British Isles. For the use of Beginners and Amateurs. By GEORGE BENTHAM, F.R.S. 4th Edition, revised, Crown 8vo, 12s.

Illustrations of the British Flora; a Series of Wood Engravings, with Dissections, of British Plants, from Drawings by W. H. FITCH, F.L.S., and W. G. SMITH, F.L.S., forming an Illustrated Companion to BENTHAM's "Handbook," and other British Floras. 1306 Wood Engravings, 12s.

Domestic Botany; an Exposition of the Structure and Classification of Plants, and of their uses for Food, Clothing, Medicine, and Manufacturing Purposes. By JOHN SMITH, A.L.S., ex-Curator of the Royal Gardens, Kew. Crown 8vo, 16 Coloured Plates and Wood Engravings, 16s.

British Wild Flowers, Familiarly Described in the Four Seasons. By THOMAS MOORE, F.L.S. 24 Coloured Plates, 16s.

The Narcissus, its History and Culture, with Coloured Figures of all known Species and Principal Varieties. By F. W. BURBIDGE, and a Review of the Classification by J. G. BAKER, F.L.S. Super-royal 8vo, 48 Coloured Plates, 32s.

The Botanical Magazine; Figures and Descriptions
of New and Rare Plants of interest to the Botanical Student, and suitable for the Garden, Stove, or Greenhouse. By Sir J. D. HOOKER, K.C.S.I., C.B., F.R.S., Director of the Royal Gardens, Kew. Royal 8vo. Third Series, Vols. I. to XXXVIII., each 42s. Published Monthly, with 6 Plates, 3s. 6d., coloured. Annual Subscription, 42s.

RE-ISSUE of the THIRD SERIES in Monthly Vols., 42s. each; to Subscribers for the entire Series, 36s. each.

The Floral Magazine; New Series, Enlarged to
Royal 4to. Figures and Descriptions of the choicest New Flowers for the Garden, Stove, or Conservatory. Complete in Ten Vols., in handsome cloth, gilt edges, 42s. each.

FIRST SERIES complete in Ten Vols., with 560 beautifully-coloured Plates, £18 7s. 6d.

Wild Flowers of the Undercliff, Isle of Wight.
By CHARLOTTE O'BRIEN and C. PARKINSON. Crown 8vo, 8 Coloured Plates, 7s. 6d.

The Young Collector's Handybook of Botany.
By the Rev. H. P. DUNSTER, M.A. 66 Wood Engravings, 3s. 6d.

Laws of Botanical Nomenclature adopted by
the International Botanical Congress, with an Historical Introduction and a Commentary. By ALPHONSE DE CANDOLLE. 2s. 6d.

Contributions to the Flora of Mentone, and to a
Winter Flora of the Riviera, including the Coast from Marseilles to Genoa. By J. TRAHERNE MOGGRIDGE, F.L.S. Royal 8vo. Complete in One Vol., with 99 Coloured Plates, 63s.

Flora Vitiensis; a Description of the Plants of
the Viti or Fiji Islands, with an Account of their History, Uses, and Properties. By Dr. BERTHOLD SEEMANN, F.L.S. Royal 4to, Coloured Plates. Part X., 25s.

Flora of Mauritius and the Seychelles; a Description of the Flowering Plants and Ferns of those Islands. By J. G. BAKER, F.L.S. 24s. Published under the authority of the Colonial Government of Mauritius.

Flora of British India. By Sir J. D. HOOKER, K.C.S.I., C.B., F.R.S., &c.; assisted by various Botanists. Parts I. to X., 10s. 6d. each. Vols. I. to III., cloth, 32s. each. Published under the authority of the Secretary of State for India in Council.

Flora of Tropical Africa. By DANIEL OLIVER, F.R.S., F.L.S. Vols. I. to III., 20s. each. Published under the authority of the First Commissioner of Her Majesty's Works.

Handbook of the New Zealand Flora; a Systematic Description of the Native Plants of New Zealand, and the Chatham, Kermadec's, Lord Auckland's, Campbell's, and Macquarrie's Islands. By Sir J. D. HOOKER, K.C.S.I., F.R.S. Part II., CRYPTOGAMIA, 14s. Published under the auspices of the Government of that Colony.

Flora Australiensis; a Description of the Plants of the Australian Territory. By GEORGE BENTHAM, F.R.S., assisted by FERDINAND MUELLER, F.R.S., Government Botanist, Melbourne, Victoria. Complete in Seven Vols., £7 4s. Vols. I. to VI., 20s. each; Vol. VII., 24s. Published under the auspices of the several Governments of Australia.

Flora of the British West Indian Islands. By Dr. GRISEBACH, F.L.S. 42s. Published under the auspices of the Secretary of State for the Colonies.

Flora Hongkongensis; a Description of the Flowering Plants and Ferns of the Island of Hongkong. By GEORGE BENTHAM, F.R.S. With a Map of the Island, and a Supplement by Dr. HANCE. 18s. Published under the authority of Her Majesty's Secretary of State for the Colonies. The Supplement separately, 2s. 6d.

Flora Capensis; a Systematic Description of the Plants of the Cape Colony, Caffraria, and Port Natal. By WILLIAM H. HARVEY, M.D., F.R.S., Professor of Botany in the University of Dublin, and OTTO WILHEM SONDER, Ph.D. Vols. I. and II., 12s. each. Vol. III., 18s.

Flora of Hampshire, including the Isle of Wight, with localities of the less common species. By FREDERICK TOWNSEND, M.A., F.L.S. With Coloured Map and two Plates, 16s.

On the Flora of Australia: its Origin, Affinities, and Distribution; being an Introductory Essay to the "Flora of Tasmania." By Sir J. D. HOOKER, F.R.S. 10s.

Genera Plantarum, ad Exemplaria imprimis in Herbariis Kewensibus servata definita. By GEORGE BENTHAM, F.R.S., F.L.S., and Sir J. D. HOOKER, F.R.S., Director of the Royal Gardens, Kew. Complete in Three Vols., £8 5s.; or separately:—Vol. I.—Part I., Royal 8vo, 21s.; Part II., 14s; Part III., 15s.; or Vol. I. complete, 50s. Vol. II.—Part I., 24s.; Part II., 32s.; or Vol. II. complete, 56s. Vol. III.—Part I., 24s.; Part II., 36s.; or Vol. III. complete, 56s.

Illustrations of the Nueva Quinologia of Pavon, with Observations on the Barks described. By J. E. HOWARD, F.L.S. With 27 Coloured Plates. Imperial folio, half-morocco gilt edges, £6 6s.

The Quinology of the East Indian Plantations. By J. E. HOWARD, F.L.S. Complete in One Vol., folio. With 13 Coloured and 2 Plain Plates, and 2 Photo-prints, 84s. Parts II. and III., cloth, 63s.

Revision of the Natural Order Hederaceæ; being a reprint, with numerous additions and corrections, of a series of papers published in the "Journal of Botany, British and Foreign." By BERTHOLD SEEMANN, Ph.D., F.L.S. 7 Plates, 10s. 6d.

Icones Plantarum. Figures, with Brief Descriptive Characters and Remarks, of New and Rare Plants, selected from the Author's Herbarium. By Sir W. J. HOOKER, F.R.S. New Series, Vol. V. 100 Plates, 31s. 6d.

Orchids; and How to Grow them in India and
other Tropical Climates. By SAMUEL JENNINGS, F.L.S., F.R.H.S.,
late Vice-President of the Agri-Horticultural Society of India.
Royal 4to. Complete in One Vol., cloth, gilt edges, 63s.

A Second Century of Orchidaceous Plants, selected
from the Subjects published in Curtis's "Botanical Magazine"
since the issue of the "First Century." Edited by JAMES BATE-
MAN, Esq., F.R.S. Complete in One Vol., Royal 4to, 100 Coloured
Plates, £5 5s.

Dedicated by Special Permission to H.R.H. the Princess of Wales.
Monograph of Odontoglossum, a Genus of the
Vandeous Section of Orchidaceous Plants. By JAMES BATEMAN,
Esq., F.R.S. Imperial folio, complete in Six Parts, each with 5
Coloured Plates, and occasional Wood Engravings, 21s.; or, in
One Vol., cloth, £6 16s. 6d.

The Rhododendrons of Sikkim-Himalaya; being
an Account, Botanical and Geographical, of the Rhododendrons
recently discovered in the Mountains of Eastern Himalaya, by
Sir J. D. Hooker, F.R.S. By Sir W. J. HOOKER, F.R.S. Folio,
30 Coloured Plates, £4 14s. 6d.

Outlines of Elementary Botany, as Introductory
to Local Floras. By GEORGE BENTHAM, F.R.S., President of
the Linnæan Society. New Edition, 1s.

British Grasses; an Introduction to the Study
of the Gramineæ of Great Britain and Ireland. By M. PLUES.
Crown 8vo, with 16 Coloured Plates and 100 Wood Engravings,
10s. 6d.

Familiar Indian Flowers. By LENA LOWIS. 4to,
30 Coloured Plates, 31s. 6d.

Botanical Names for English Readers. By RANDAL
H. ALCOCK. 8vo, 6s.

Elementary Lessons in Botanical Geography. By
J. G. BAKER, F.L.S. 3s.

FERNS.

British Ferns; an Introduction to the Study of
the FERNS, LYCOPODS, and EQUISETA indigenous to the British Isles. With Chapters on the Structure, Propagation, Cultivation, Diseases, Uses, Preservation, and Distribution of Ferns. By M. PLUES. Crown 8vo, with 16 Coloured Plates, and 55 Wood Engravings, 10s. 6d.

The British Ferns; Coloured Figures and Descrip-
tions, with Analysis of the Fructification and Venation of the Ferns of Great Britain and Ireland. By Sir W. J. HOOKER, F.R.S. Royal 8vo, 66 Coloured Plates, £2 2s.

Garden Ferns; Coloured Figures and Descriptions
with Analysis of the Fructification and Venation of a Selection of Exotic Ferns, adapted for Cultivation in the Garden, Hothouse, and Conservatory. By Sir W. J. HOOKER, F.R.S. Royal 8vo, 64 Coloured Plates, £2 2s.

Filices Exoticæ; Coloured Figures and Description
of Exotic Ferns. By Sir W. J. HOOKER, F.R.S. Royal 4to, 100 Coloured Plates, £6 11s.

Ferny Combes; a Ramble after Ferns in the Glens
and Valleys of Devonshire. By CHARLOTTE CHANTER. Third Edition. Fcap. 8vo, 8 Coloured Plates and a Map of the County, 5s.

MOSSES.

Handbook of British Mosses, containing all that
are known to be natives of the British Isles. By the Rev. M. J. BERKELEY, M.A., F.L.S. 24 Coloured Plates, 21s.

Synopsis of British Mosses, containing Descrip-
tions of all the Genera and Species (with localities of the rarer ones) found in Great Britain and Ireland. By CHARLES P. HOBKIRK, President of the Huddersfield Naturalist's Society. Crown 8vo, 7s. 6d.

SEAWEEDS.

British Seaweeds; an Introduction to the Study of
the Marine ALGÆ of Great Britain, Ireland, and the Channel Islands. By S. O. GRAY. Crown 8vo, with 16 Coloured Plates, 10s. 6d.

Phycologia Britannica; or, History of British
Seaweeds. Containing Coloured Figures, Generic and Specific Characters, Synonyms and Descriptions of all the Species of Algæ inhabiting the Shores of the British Islands. By Dr. W. H. HARVEY, F.R.S. New Edition. Royal 8vo, 4 vols. 360 Coloured Plates, £7 10s.

Phycologia Australica; a History of Australian
Seaweeds, comprising Coloured Figures and Descriptions of the more characteristic Marine Algæ of New South Wales, Victoria, Tasmania, South Australia, and Western Australia, and a Synopsis of all known Australian Algæ. By Dr. W. H. HARVEY, F.R.S. Royal 8vo, Five Vols., 300 Coloured Plates, £7 13s.

FUNGI.

Outlines of British Fungology, containing Cha-
racters of above a Thousand Species of Fungi, and a Complete List of all that have been described as Natives of the British Isles. By the Rev. M. J. BERKELEY, M.A., F.L.S. 24 Coloured Plates, 30s.

The Esculent Funguses of England. Containing
an Account of their Classical History, Uses, Characters, Development, Structure, Nutritious Properties, Modes of Cooking and Preserving, &c. By C. D. BADHAM, M.D. Second Edition. Edited by F. CURREY, F.R.S. 12 Coloured Plates, 12s.

Clavis Agaricinorum; an Analytical Key to the
British Agaricini, with Characters of the Genera and Sub-genera. By WORTHINGTON G. SMITH, F.L.S. 6 Plates, 2s. 6d.

SHELLS AND MOLLUSKS.

Testacea Atlantica; or, the Land and Freshwater
Shells of the Azores, Madeiras, Salvages, Canaries, Cape Verdes, and Saint Helena. By T. VERNON WOLLASTON, M.A., F.L.S. Demy 8vo, 25s.

Elements of Conchology; an Introduction to
the Natural History of Shells, and of the Animals which form them. By LOVELL REEVE, F.L.S. Royal 8vo, Two Vols., 62 Coloured Plates, £2 16s.

Conchologia Iconica; or, Figures and Descriptions
of the Shells of Mollusks, with remarks on their Affinities, Synonymy, and Geographical Distribution. By LOVELL REEVE, F.L.S., and G. B. SOWERBY, F.L.S., complete in Twenty Vols., 4to, with 2727 Coloured Plates, half-calf, £178.

A detailed list of Monographs and Volumes may be had.

Conchologia Indica; Illustrations of the Land and
Freshwater Shells of British India. Edited by SYLVANUS HANLEY, F.L.S., and WILLIAM THEOBALD, of the Geological Survey of India. Complete in One Vol., 4to, with 160 Coloured Plates, £8 5s.

The Edible Mollusks of Great Britain and Ireland,
with the Modes of Cooking them. By M. S. LOVELL. Crown 8vo, with 12 Coloured Plates, 8s. 6d.

INSECTS.

The Lepidoptera of Ceylon. By F. MOORE, F.L.S.
Parts I. to VII. Medium 4to, each with 18 Plates, to Subscribers only, 31s. 6d., coloured; 16s., uncoloured. Also Vol. I. (Rhopalocera), complete, cloth, gilt top, £7 15s.; to Subscribers for the entire work, £6 10s. Published under the auspices of the Government of Ceylon.

The Butterflies of Europe; Illustrated and Described. By HENRY CHARLES LANG, M.D., F.L.S. Super-royal 8vo, Parts I. to XIII., each, with 4 Coloured Plates, 3s. 6d. To be completed in about 20 Parts.

The Larvæ of the British Lepidoptera, and their Food Plants. By OWEN S. WILSON. With Life-size Figures, drawn and coloured from Nature by ELEANORA WILSON. Super-royal 8vo. With 40 elaborately-coloured Plates, containing upwards of 600 figures of Larvæ, 63s.

British Insects. A Familiar Description of the Form, Structure, Habits, and Transformations of Insects. By E. F. STAVELEY, Author of "British Spiders." Crown 8vo, with 16 Coloured Plates and numerous Wood Engravings, 14s.

British Beetles; an Introduction to the Study of our indigenous COLEOPTERA. By E. C. RYE. Crown 8vo 16 Coloured Steel Plates, and 11 Wood Engravings, 10s. 6d.

British Bees; an Introduction to the Study of the Natural History and Economy of the Bees indigenous to the British Isles. By W. E. SHUCKARD. Crown 8vo, 16 Coloured Plates, and Woodcuts of Dissections, 10s. 6d.

British Butterflies and Moths; an Introduction to the Study of our Native LEPIDOPTERA. By H. T. STAINTON. Crown 8vo, 16 Coloured Plates, and Wood Engravings, 10s. 6d.

British Spiders; an Introduction to the Study of the ARANEIDÆ found in Great Britain and Ireland. By E. F. STAVELEY. Crown 8vo, 16 Coloured Plates, and 44 Wood Engravings, 10s. 6d.

Harvesting Ants and Trap-door Spiders; Notes and Observations on their Habits and Dwellings. By J. T. MOGGRIDGE, F.L.S. With a SUPPLEMENT of 160 pp. and 8 additional Plates, 17s. The Supplement separately, cloth, 7s. 6d.

Curtis's British Entomology. Illustrations and
Descriptions of the Genera of Insects found in Great Britain
and Ireland, Containing Coloured Figures, from Nature, of the
most rare and beautiful Species, and in many instances, upon the
plants on which they are found. Eight Vols., Royal 8vo, 770
Coloured Plates, £28.

Or in Separate Monographs.

Orders.	Plates.	£ s. d.	Orders.	Plates.	£ s. d.
APHANIPTERA	2	0 2 0	HYMENOPTERA	125	6 5 0
COLEOPTERA	256	12 16 0	LEPIDOPTERA	193	9 13 0
DERMAPTERA	1	0 1 0	NEUROPTERA	13	0 13 0
DICTYOPTERA	1	0 1 0	OMALOPTERA	6	0 6 0
DIPTERA	103	5 3 0	ORTHOPTERA	5	0 5 0
HEMIPTERA	32	1 12 0	STREPSIPTERA	3	0 3 0
HOMOPTERA	21	1 1 0	TRICHOPTERA	9	0 9 0

"Curtis's Entomology," which Cuvier pronounced to have "reached the ultimatum of perfection," is still the standard work on the Genera of British Insects. The Figures executed by the author himself, with wonderful minuteness and accuracy, have never been surpassed, even if equalled. The price at which the work was originally published was £43 16s.

Insecta Britannica; Vol. III., Diptera. By
FRANCIS WALKER, F.L.S. 8vo, with 10 Plates, 25s.

ANTIQUARIAN.

Sacred Archæology; a Popular Dictionary of
Ecclesiastical Art and Institutions from Primitive to Modern
Times. Comprising Architecture, Music, Vestments, Furniture
Arrangement, Offices, Customs, Ritual Symbolism, Ceremonial
Traditions, Religious Orders, &c., of the Church Catholic in all
ages. By MACKENZIE E. C. WALCOTT, B.D. Oxon., F.S.A.,
Precentor and Prebendary of Chichester Cathedral. Demy 8vo,
18s.

A Manual of British Archæology. By CHARLES
BOUTELL, M.A. 20 Coloured Plates, 10s. 6d.

The Antiquity of Man; an Examination of Sir
Charles Lyell's recent Work. By S. R. PATTISON, F.G.S.
Second Edition. 8vo, 1s.

MISCELLANEOUS.

Handbook of the Vertebrate Fauna of Yorkshire;
being a Catalogue of Mammals, Birds, Reptiles, Amphibians,
and Fishes, which are or have been found in the County. By
W. E. CLARKE and W. D. ROEBUCK. 8vo, 8s. 6d.

Report on the Forest Resources of Western
Australia. By Baron FERD. MUELLER, C.M.G., M.D., Ph.D.,
F.R.S., Government Botanist of Victoria. Royal 4to, 20
Plates, 12s.

West Yorkshire; an Account of its Geology, Physical
Geography, Climatology, and Botany. By J. W. DAVIS, F.L.S.,
and F. ARNOLD LEES, F.L.S. Second Edition, 8vo, 21 Plates,
many Coloured, and 2 large Maps, 21s.

Handbook of the Freshwater Fishes of India;
giving the Characteristic Peculiarities of all the Species at
present known, and intended as a guide to Students and District
Officers. By Capt. R. BEAVAN, F.R.G.S. Demy 8vo, 12 plates,
10s. 6d.

Natal; a History and Description of the Colony,
including its Natural Features, Productions, Industrial Condition
and Prospects. By HENRY BROOKS, for many years a resident.
Edited by Dr. R. J. MANN, F.R.A.S., F.R.G.S., late Superintendent of Education in the Colony. Demy 8vo, with Maps,
Coloured Plates, and Photographic Views, 21s.

St. Helena. A Physical, Historical, and Topographical Description of the Island, including its Geology, Fauna,
Flora, and Meteorology. By J. C. MELLISS, A.I.C.E., F.G S.,
F.L.S. In one large Vol., Super-royal 8vo, with 56 Plates and
Maps, mostly coloured, 42s.

Lahore to Yarkand. Incidents of the Route and
Natural History of the Countries traversed by the Expedition of

1870, under T. D. FORSYTH, Esq., C.B. By GEORGE HENDERSON, M.D., F.L.S., F.R.G.S., and ALLAN O. HUME, Esq., C.B., F.Z.S. With 32 Coloured Plates of Birds, 6 of Plants, 26 Photographic Views, Map, and Geological Sections, 42s.

The Birds of Sherwood Forest; with Observations on their Nesting, Habits, and Migrations. By W. J. STERLAND. Crown 8vo, 4 plates. 7s. 6d., coloured.

The Young Collector's Handy Book of Recreative Science. By the Rev. H. P. DUNSTER, M.A. Cuts, 3s. 6d.

A Survey of the Early Geography of Western Europe, as connected with the First Inhabitants of Britain, their Origin, Language, Religious Rites, and Edifices. By HENRY LAWES LONG, Esq. 8vo, 6s.

The Geologist. A Magazine of Geology, Palæontology, and Mineralogy. Illustrated with highly-finished Wood Engravings. Edited by S. J. MACKIE, F.G.S., F.S.A. Vols. V. and VI., each with numerous Wood Engravings, 18s. Vol. VII., 9s.

Everybody's Weather-Guide. The use of Meteorological Instruments clearly explained, with directions for securing at any time a probable Prognostic of the Weather. By A. STEINMETZ, Esq., Author of " Sunshine and Showers," &c. 1s.

The Artificial Production of Fish. By PISCARIUS. Third Edition. 1s.

The Gladiolus: its History, Cultivation, and Exhibition. By the Rev. H. HONYWOOD DOMBRAIN, B.A. 1s.

Meteors, Aerolites, and Falling Stars. By Dr. T. L. PHIPSON, F.C.S. Crown 8vo, 25 Woodcuts and Lithographic Frontispiece, 6s.

The Zoology of the Voyage of H.M.S. *Samarang*,
under the command of Captain Sir Edward Belcher, C.B., during
the Years 1843-46. By Professor OWEN, Dr. J. E. GRAY, Sir J.
RICHARDSON, A. ADAMS, L. REEVE, and A. WHITE. Edited by
ARTHUR ADAMS, F.L.S. Royal 4to, 55 Plates, mostly coloured,
£3 10s.

Papers for the People. By ONE OF THEM. No. 1,
Our Land. 8vo, 6d. (By Post, 7d. in stamps.)

The Royal Academy Album; a Series of Photographs from Works of Art in the Exhibition of the Royal Academy
of Arts, 1875. Atlas 4to, with 32 fine Photographs, cloth,
gilt edges, £6 6s.; half-morocco, £7 7s.

The same for 1876, with 48 beautiful Photo-prints, cloth,
£6 6s.; half-morocco, £7 7s. Small Edition, Royal 4to, cloth, gilt
edges, 63s.

On Intelligence. By H. TAINE, D.C.L. Oxon.
Translated from the French by T. D. HAYE, and revised, with
additions, by the Author. Complete in One Vol., 18s.

Manual of Chemical Analysis, Qualitative and
Quantitative; for the use of Students. By Dr. HENRY M. NOAD,
F.R.S. New Edition. Crown 8vo, 109 Wood Engravings, 16s.
Or, separately, Part I., "QUALITATIVE," New Edition, new
Notation, 6s.; Part II., "QUANTITATIVE," 10s. 6d.

Live Coals; or, Faces from the Fire. By L.
M. BUDGEN, "Acheta," Author of "Episodes of Insect Life,"
&c. Dedicated, by Special Permission, to H.R.H. Field Marshal
the Duke of Cambridge. Royal 4to, 35 Original Sketches printed
in colours, 21s.

Caliphs and Sultans; being tales omitted in the
ordinary English version of "The Arabian Nights' Entertainments," freely rewritten and rearranged. By S. HANLEY, F.L.S.
6s.

PLATES.

Floral Plates, from the Floral Magazine. Beautifully Coloured, for Screens, Scrap-books, Studies in Flower-painting, &c. 6d. and 1s. each. Lists of over 700 varieties, One Stamp.

Botanical Plates, from the Botanical Magazine. Beautifully-coloured Figures of new and rare Plants. 6d. and 1s. each. Lists of over 2000, Three Stamps.

SERIALS.

The Botanical Magazine. Figures and Descriptions of New and rare Plants. By Sir J. D. HOOKER, C.B., F.R.S. Monthly, with 6 Coloured Plates, 3s. 6d. Annual subscription, post free, 42s.

Re-issue of the Third Series, in Monthly Vols., 42s. each; to Subscribers for the entire Series, 36s. each.

The Lepidoptera of Ceylon. By F. MOORE. 16s., plain; 31s. 6d., coloured.

The Butterflies of Europe. By Dr. LANG, F.L.S. Monthly, with 4 Coloured Plates, 3s. 6d. Subscription for the complete Work (20 Parts), in advance, 60s.

FORTHCOMING WORKS.

Flora of India. By Sir J. D. HOOKER and others. Part XI.

Natural History of Plants. By Prof. BAILLON.

Flora of Tropical Africa. By Prof. OLIVER.

Flora Capensis. By Prof. DYER.

London:
L. REEVE & CO., 5, HENRIETTA STREET, COVENT GARDEN.

PRINTED BY GILBERT AND RIVINGTON, LIMITED, ST. JOHN'S SQUARE, LONDON.